AF809235

Management for Professionals

The Springer series *Management for Professionals* comprises high-level business and management books for executives. The authors are experienced business professionals and renowned professors who combine scientific background, best practice, and entrepreneurial vision to provide powerful insights into how to achieve business excellence.

More information about this series at http://www.springer.com/series/10101

Reiner Hutwelker

Six Sigma Green Belt Certification Project

Identification, Implementation and Evaluation

Springer

Reiner Hutwelker
SigmaLogic
Aich, Austria

ISSN 2192-8096 ISSN 2192-810X (electronic)
Management for Professionals
ISBN 978-3-030-31914-4 ISBN 978-3-030-31915-1 (eBook)
https://doi.org/10.1007/978-3-030-31915-1

© Springer Nature Switzerland AG 2019
This work is subject to copyright. All rights are reserved by the Publisher, whether the whole or part of the material is concerned, specifically the rights of translation, reprinting, reuse of illustrations, recitation, broadcasting, reproduction on microfilms or in any other physical way, and transmission or information storage and retrieval, electronic adaptation, computer software, or by similar or dissimilar methodology now known or hereafter developed.
The use of general descriptive names, registered names, trademarks, service marks, etc. in this publication does not imply, even in the absence of a specific statement, that such names are exempt from the relevant protective laws and regulations and therefore free for general use.
The publisher, the authors, and the editors are safe to assume that the advice and information in this book are believed to be true and accurate at the date of publication. Neither the publisher nor the authors or the editors give a warranty, expressed or implied, with respect to the material contained herein or for any errors or omissions that may have been made. The publisher remains neutral with regard to jurisdictional claims in published maps and institutional affiliations.

This Springer imprint is published by the registered company Springer Nature Switzerland AG.
The registered company address is: Gewerbestrasse 11, 6330 Cham, Switzerland

And this is for you, Nora*: I know you are there, somewhere – because your absence fills my life . . . as a star in the night and the light for my day. And one day you will understand, what was done to you.*

Preface

You want to realize your own Six Sigma project? Here is a sample project about the low quality, the lack of availability, and the consumption-intensive production of cookies in a bakery. From the identification of a suitable project topic in the DEFINE to the sustainable management of its success in the CONTROL phase, all necessary steps are demonstrated and supported by a DMAIC software guideline. The application of the sequentially linked DMAIC tools is easy to understand and directly transferable to typical individual Six Sigma Business Projects for professionals. As a student, still in education, you can implement our predefined Standard Project on environmental littering, based on concrete tasks. Both the Business and the Standard Project can lead to your Green Belt certification. Test questions and answers can stabilize your knowledge. Tips for your project Sponsor clarify this management role and provide basic knowledge of Six Sigma, checklists, and information on success factors and threats for Six Sigma projects and enterprise programs. Notes for Six Sigma experts explain the concept and the selection, sequencing, and modification of the tools.

Learning is supported by:

- Videos: Animated content of this book (43 lessons, 5:15 hours)
- sigmaGuide: Software for the implementation of Six Sigma projects (32 tools and charts)
- Methods and tools for process improvement (86 slides with the figures of this book)
- Statistics with Minitab (136 slides)
- Project-Story-Book: PowerPoint template for project documentation (60 slides)
- Project Guideline: Overview poster of all DMAIC phases and tools

The course takes place in partnership with the TU Munich, as part of the edX Professional Series:

- *Six Sigma and Lean: Quantitative Tools for Quality and Productivity (QPLSx)*: www.edx.org.
- The certification is awarded by the TU München School of Management: www.eec.wi.tum.de.

The handouts and the software are additionally available on my homepage: www.sigmalogic.de and the videos on my YouTube channel: https://www.youtube.com/channel/UCShsy50l5dFsoSXU__hBKSA.

All illustrations are by the author, except Figs. 2.3 and 13.1 and Table 6.1.

Aich, Austria
August 2019

Reiner Hutwelker

Acknowledgements

Thanks to Martin Grunow from TU Munich and Bernhard Kraus from the TUM School of Management for their trust and opportunities. My very special and heartfelt thanks to Holly Ott from TUM. She has built the ship, set the sails, is its helmsman, and is its brain, heart, and soul. And she provides the wind.

Contents

About the Author

Reiner Hutwelker Six Sigma Master Black Belt, formerly Siemens Management Consulting, today consultant, trainer, and coach for process improvements by Six Sigma and Lean. With more than 25 years of experience in production and services, Dr. Hutwelker specializes in the development of methods (business game: codierFabrik) and tools (software: sigmaGuide) and is Senior Lecturer at TU Munich and TUM School of Management (EEC).

Certification Path, Projects, and Course Concept

1

Hello, my name is Reiner Hutwelker and I am a Six Sigma Master Black Belt and Senior Lecturer at Technical University Munich (TUM). This (Lean) Sigma Green Belt Certification Course is the follow-up course to the TUM Yellow Belt. The course is about the implementation of a concrete Six Sigma project. This can be a Business-Project with an individual topic of your company or our Standard-Project about environmental littering in your hometown, if you are still in education. In each of these projects, you will acquire the skills to apply the Six Sigma methodology and tools and to transfer them to other typical problems in production and service. You should like to observe, analyze facts, and draw the appropriate conclusions, because that is what we will do here.

1.1 Certification Requirements

There are many different offers on the net for the Green Belt certificates and certification. Wikipedia summarizes: "Criteria for Green Belt and Black Belt certification vary . . . There is no standard certification body" (https://en.wikipedia.org/wiki/Six_Sigma#Certification). For your orientation we compared our requirements with those of the ASQ, the American Society for Quality (Table 1.1).

The examination of the Six Sigma theory knowledge is a necessary condition for all providers. The ASQ has set a standard with its body of knowledge (BoK). This BoK is the reference for us. It is taught and tested by our TUM Lean Six Sigma Yellow Belt program (TUMx—QPLS1-3x). The prerequisite for participation in this Green Belt Course is that you already acquired the TUM Yellow Belt certificate. And please note: We do not accept Yellow Belt certificates from other providers.

The ASQ requires 3 years of professional experience in process improvement under the guidance of a Black Belt. This requirement, however, not only excludes

Video 1 (https://www.youtube.com/channel/UCShsy50l5dFsoSXU__hBKSA)

© Springer Nature Switzerland AG 2019

R. Hutwelker, *Six Sigma Green Belt Certification Project*, Management for Professionals, https://doi.org/10.1007/978-3-030-31915-1_1

Table 1.1 Comparison of the certification requirements of ASQ and TUM for the Green Belt

		Examination of BoK [4]?	Work experience required?	Certification Project?	Coaching?		Type of Certificate?	Direct Benefit?
					Open Online Sessions	Individual Coaching		
ASQ [1]		yes	3 years, full time, paid role; in one or more areas of the GB BoK; support/ under supervision of a BB; involved in CIP [6]	no			ASQ	participant
edX	TUM [2]	yes [5]	no	yes - predefined Standard-Project - implemented by participant - reviewed by MBB	yes	no	TUM Six Sigma Green Belt	participant
edX	TUM/ EEC [3]	yes [5]	no	yes - individual Business-Project - implemented by participant - reviewed by MBB	yes	yes	TUM Lean Six Sigma Green Belt	participant & company

1) American Society for Quality
2) Technical University of Munich
3) TUM Executive Education Center
4) Body of Knowledge
5) TUM Lean and Six Sigma Yellow Belt Certification
6) Source: ASQ -Certified Six Sigma Green Belt

new employees, but all employees of companies without Six Sigma initiatives or Black Belts. We do not require prior work experience.

In contrast to us, ASQ does not require the implementation of a certification project. The ISSSP (International Society of Six Sigma Professionals) distinguishes between a *certificate* awarded for the theory examination and a *certification* for the completion of a project. This is exactly how we see it: After passing a theory exam you cannot drive a car yet—you need driving lessons. Therefore, we demand the implementation of a concrete project for our Green Belt certification. Our course and this book are meant to support you.

1.2 Two Types of Certification Projects: For Professionals and Students

Typically, a certification project is carried out based on a concrete problem in a company. We offer the certification of professionals based on these individual company-specific **Business-Projects** and we support them by individual coaching and reviews. First, we support you in identifying a suitable project topic and in defining the project. Depending on its scope, this project can lead to the *Six Sigma* or to the *Lean Six Sigma Green Belt Certification* (Fig. 1.1).

We also want to open our course to students. Our **Standard-Project** deals with the littering of public places in your hometown and is, therefore, already identified and predefined: *Improve the cleanliness of areas around selected places—e.g., public trash cans—and control the sustainability of your measures.* This topic

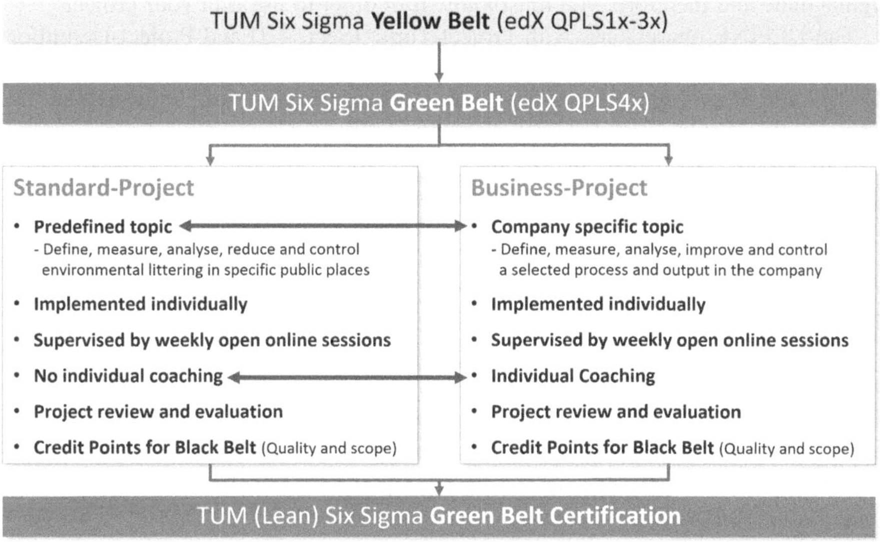

Fig. 1.1 Two ways to the Six Sigma Green Belt: one for students by implementing our Standard-Project and one for professionals by implementing an individual Business-Project in a company

concerns everyone: You get annoyed about it or contribute to it yourself. Here a paper cup, a plastic lid and a straw, there a burger box and finally there might be a cigarette butt or a chewing gum. Of course we want to get rid of the garbage right away and sometimes the *coffee to go* becomes a *cup to throw*.

Many do this, men more often than women, younger people more often than older people. It happens in the countryside and in the cities, on the streets, at bus stops, on parking lots, near restaurants and in parks, although there is often a trash can very close by. We would like to improve this situation a little with you in this Standard-Project. It is suitable as a certification project because you will be able to use all the core DMAIC tools, from the project definition to data collection, test of hypotheses, root cause analysis, implementation and control of solutions.

The milestones of Business- and Standard-Projects are of course based on the DMAIC phases (Table 1.2). The course content is explained in the videos, in the corresponding handouts and in further handouts for the use of *sigmaGuide*, for the use of the *Project-Story-Book* and *Minitab*.

sigmaGuide is a software guide with all important rational tools for your project. Our *Project-Story-Book* template is necessary to document the results of the phases. And *Minitab* is a statistical software, available as a free 30-day trial version. Of course, you can use any software that allows you to perform the necessary data analysis with statistical tests and control charts (e.g., gretl, R).

Please make sure that you work through the material carefully because we will check your knowledge after each chapter with questions. And we provide you with specific tasks for each step.

In the Introduction (Chap. 2), we will review some basics about Six Sigma and present an integrated process-problem-solving model. This model is the basis of the sigmaGuide and therefore, you must know it in order to use it in your project.

The DEFINE phase starts with Project-Topic (Sect. 4.2) and Project-Definition (Sect. 4.3). For the Standard-Project we already predefined some information in the sigmaGuide for your orientation. In Business-Projects, all information for sigmaGuide must be collected from a specific business process.

You should then copy the results of both tools into the Project-Story-Book template (Chap. 5), comment on them concisely, and supplement them with a clear illustration of the defined problems. Together with some information about you, supplemented by a scan of your TUM Yellow Belt certificate and for Business-Projects additionally about your Sponsor and your company, you will have documented the first milestone of your project.

We support both project types by open online sessions on our digital learning platform. Candidates with Business-Projects are additionally supported by individual online coaching. And together with a responsible manager, your project Sponsor, we decide on the successful completion of each phase. An introduction to Six Sigma for Sponsors, checklists and analyses of Six Sigma programs in German companies support the responsible manager in assuming the role of your project Sponsor (Chap. 14). If you are running a Standard-Project, you can proceed from phase to phase without individual coaching and without our decisions.

Table 1.2 Certification milestones (download)

Six Sigma Green Belt Certification Process	Introduction	DEFINE (part 1 & 2)	DEFINE (part 3)	MEASURE	ANALYSE	IMPROVE	CONTROL
Course Content	Videos	Videos, Handouts, sigmaGuide, Project-Story-Book	Videos, Handouts, sigmaGuide, Project-Story-Book	Videos, Handouts, sigmaGuide, Project-Story-Book	Videos, Handouts, sigmaGuide, Project-Story-Book	Videos, Handouts, sigmaGuide, Project-Story-Book	Videos, Handouts, sigmaGuide, Project-Story-Book
	1 Certification-Path 2 Course Introduction 3 Six Sigma Introduction: Overview 4 Six Sigma Introduction: Improvement Program 5 Six Sigma Example: Cookie du Chef 6 Six Sigma Introduction: Method DMAIC 7 Six Sigma Introduction: Method Process-Problem-Solving-Model 8 Six Sigma Introduction: Method Problem-Solving-Principles 9 Six Sigma Introduction: Method Statistical-Basis 10 Six Sigma Introduction: Linking Reality Model & Statistics	11 sigmaGuide 12 DEFINE Overview 13 DEFINE Project-Topic 14 DEFINE Project-Definition 15 Project-Story-Book	16 DEFINE SIPOC 17 DEFINE Voice-to-Critical 18 DEFINE Project-Charter 19 DEFINE Stakeholder-Communication	20 MEASURE Overview 21 MEASURE Input-Analysis 22 MEASURE Process-Mapping-Analysis 23 MEASURE C&E-Matrix 24 MEASURE Data-Collection-Plan 25 MEASURE Hypotheses	26 ANALYSE Overview 27 ANALYSE Data-Evaluation 28 ANALYSE Processs-Performance 29 ANALYSE Hypotheses-Tests 30 ANALYSE Root-Cause-Analysis - Preparation 31 ANALYSE Root-Cause-Analysis - Implementation	32 IMPROVE Overview 33 IMPROVE Development and Selection of Solutions 34 IMPROVE Measures, Risks (FMEA) and Implementation	35 CONTROL Overview 36 CONTROL Data Evaluation 37 CONTROL Process-Performance 38 CONTROL Improvement-Verification and Benefits 39 CONTROL Process-Management-Plan 40 CONTROL Project-Completion 41 Six Sigma Project-Guideline **Addendum** 42 Lean and Six-Sigma - Roots and Development 43 Lean vs Six-Sigma - Comparison of Characteristics
Deliverables & Checks	- TUM Lean Six Sigma Yellow Belt Certificate - Exam questions	- Exam questions - sigmaGuide: Related tools - Results edited in Project-Story-Book (D----)	- Exam questions - sigmaGuide: Related tools - Interviews with local litter removal service (VoB) and pedestrians (VoC) - Results edited in Project-Story-Book (D----)	- Exam questions - sigmaGuide: Related tools - Minitab: Worksheet with collected data - Results edited in Project-Story-Book (DM---)	- Exam questions - sigmaGuide: Related tools - Minitab: Data Inspections and Analyses - Results edited in Project-Story-Book (DMA--)	- Exam questions - sigmaGuide: Related tools - Implementation of measures - Results edited in Project-Story-Book (DMAI-)	- Exam questions - sigmaGuide: Related tools - Minitab: - Process-Capability & Control-Charts; - Verification of benefits - Results edited in Project-Story-Book (DMAIC)
Coaching (Business-Projects)		- written feedback on the: suitability of the Project-Topic and quality of the Project-Definition	- written feedback - individual online sessions, scheduled when necessary				
Coaching (Standard- & Business-Projects)			- group online sessions: - introduction of a new DMAIC phase, tools and tasks; Q&A´s (weekly) - introduction of participants, presentation and discussion of projects (weekly) (of the total of approx. 40 sessions at least 10 must be attended)				
Decision about phase success (Business-Projects)		Master Black Belt (TUM)	Master Black Belt (TUM) Manager of your company as sponsor of your project				
Timeline (Example)							
Start of Phase	01.01.20xx	23.01.20xx	07.02.20xx	29.02.20xx	26.04.20xx	25.05.20xx	21.07.20xx
Suggested duration (days)	21	14	21	56	28	56	14
Suggested End of Phase	22.01.20xx	06.02.20xx	28.02.20xx	25.04.20xx	24.05.20xx	20.07.20xx	04.08.20xx

You already know, e.g., the tools SIPOC (Sect. 6.1), Voice of Customer (VoC), Voice of Business (VoB), CtQs, and the Project-Charter from the TUM Yellow Belt. For the Standard-Project you will have to conduct short interviews with a few pedestrians (VoC) and with a representative of the city hall and the local garbage disposal service (VoB) about the garbage littering situation in your hometown. In the Business-Projects you must of course interview (internal) customers (VoC) and responsible managers (VoB) (Sect. 6.2).

The information collected in sigmaGuide must be transferred again to your Project-Story-Book and commented on in order to document this milestone. Once again, we support both project types with open online sessions. In addition, Business-Projects are coached individually again and together with your Sponsor we decide on the successful completion of the DEFINE phase and on any necessary corrections (Table 1.2).

All further phases follow the same principle. Watch the videos, answer the questions, complete the tasks, summarize the collected information in sigmaGuide, transfer the results to your Project-Story-Book, comment on them, and draw conclusions. There is coaching again and a decision about the phase success for Business-Projects. With a Standard-Project you can again move to the next phase without coaching and without decisions but with the support of our regular open online sessions.

At the end of the MEASURE (Chap. 7) phase you will have formulated hypotheses and a worksheet with the collected data. In the ANALYSE phase (Chap. 8) you will test your hypotheses statistically based on your data and identify the root causes of the problems. In the IMPROVE phase (Chap. 9) you will develop and implement solutions and in the CONTROL phase (Chap. 10) you will test the success of your measures statistically based on new data and, verify the benefits and ensure the sustainability of your improvements.

A typical Green Belt project can be implemented within *half a year*. In the timeline (Table 1.2) we recommend different time spans for the milestones based on our experience. But we also know that your daily work has priority and that there are good reasons for delays. Therefore, you overall have *one year* to complete your Standard- or Business-Project. Thus, there are no consequences if you cannot complete a phase according to our recommendation.

A prerequisite for participation in the Green Belt course is your TUM Yellow Belt certificate to ensure that you have the necessary theoretical knowledge for the project. Regardless of whether you then implemented a Business- or a Standard-Project, we will review your Project-Story-Book after completing the CONTROL phase. For Standard-Projects we also expect you to send us the worksheet with your collected data (Excel file/Minitab project file). We will give you feedback and suggest corrections or adjustments to help you meet the requirements.

Six Sigma projects can vary greatly in their complexity and scope, as can be seen, for example, from the number of hypotheses examined. By specifying or checking the Project-Definition, we want to a priori ensure that your project does not become too complex. However, if you voluntarily carry out a very demanding Standard- or Business-Project, we will reward you with additional credit points. These credit points can be recognized, if you additionally want to acquire the Black Belt Certification. Your additional effort *now* would be credited to you *later*.

Passion for Six Sigma is your ticket to our certificate because you have some work to do, regardless of whether you are implementing our standard or an individual business project. But in the end, you would not only have passed the theory test with our Yellow Belt certificate, but you would also have proven with this project that you can drive a car. This Six Sigma Green Belt Certification has a high value for companies because you could immediately get into another car and drive off. And with your Project-Story-Book you have an excellently documented project that you could present anywhere, if you want to prove your competence and working style.

1.3 Course Introduction[1]

Welcome to our Green Belt course. In the introduction I will first give you an overview of Six Sigma as an enterprise-wide improvement mechanism (Sect. 2.2). Then Wanda, the three-year-old chef of our bakery: Cookie du Chef will demonstrate the business case for our example project (Sect. 2.3). Wanda will show you how to make cookies. She demonstrates the problems of the outputs (Y) and their causes in the process (x). We hope you enjoy this video. But you should not take it too seriously.

The introduction is continued with an integrated process-problem-solving model, allowing problems (Y) to be identified and distinguished from their causes (x). This model structures all DMAIC phases of your project (Sect. 2.4). Finally, the statistical basis of Six Sigma is presented, and everyday examples of this process performance indicator are given (Sect. 2.5).

The DMAIC cycle, its tools, and their application in our project example is the focus of our course.

With the first tool of sigmaGuide (Chap. 3) in the DEFINE phase we will identify the Project-Topic of our bakery (Sect. 4.2) and then define the project (Sect. 4.3). Your task now is to transfer the tool to your own project. For Business-Projects, a specific topic in the company must be identified, defined, and documented in the empty version of the sigmaGuide. For the Standard-Project on environmental littering, some data of the sigmaGuide are already predefined.

The following chapter shows you how to transfer the results of the sigmaGuide and your further insights into your Project-Story-Book (Chap. 5).

Then the DEFINE phase is continued with the SIPOC and further basic tools that you should apply to your project (Chap. 6).

For both project types you must collect data at the end of the MEASURE phase. For a Standard-Project you must collect real data on littering from your hometown and for a Business-Project real data from a process of your company (Chap. 7).

[1]Video 2 (https://www.youtube.com/channel/UCShsy50l5dFsoSXU__hBKSA)

The ANALYSE phase focuses on statistics with calculations for process capability and process control. With statistical tests and a hierarchy tree we then identify the root causes of the problems of our bakery. Your task is again to transfer the procedure to your concrete project (Chap. 8).

At the end of the IMPROVE phase (Chap. 9) you must implement concrete measures and prove their success at the end of the CONTROL phase (Chap. 10).

After you complete your project, we will review your Project-Story-Book, give you feedback, and ask you to make any necessary adjustments. And for this you will be certified by us (Chap. 11). Our certification ensures that you can transfer your acquired knowledge confidently to new topics in the service sector and production. We want to enable you to transform a raw information block into a project structure. This is the art of Six Sigma Belts.

No matter if you are just planning a Six Sigma project or if you have already completed it, the summary of the tools in the DMAIC phases (Chap. 12) can serve as a guideline for your project or to stabilize your knowledge. Your potential Sponsor might also be interested in this project overview and in the comparison and development of Six Sigma and Lean-Thinking (Chap. 13). If this manager then wants to support your project, the *Tips for the Project Sponsor* (Chap. 14) support the adoption of this important role.

Six Sigma Introduction

2

2.1 Overview: Steps, Tools, Their Purpose, and Results[1]

Six Sigma is a concept with at least three meanings:

1. William B. Smith and Mikel J. Harry (1986) developed Six Sigma at Motorola. The trigger was a specific quality problem in Motorola's semiconductor division and one customer explicitly required a performance level of 6 sigma.
2. This requirement was the impetus for the development of the DMAIC method and the origin of the name for the whole improvement program (personal message from Motorola).
3. Jack Welsh (1995) at General Electric perfected the implementation, penetration, and sustainable establishment of the improvement program in the whole organization (Fig. 2.1).

1. Excellent Six Sigma programs are enterprise-wide improvement mechanisms, to increase customer satisfaction and to reduce costs. Initially, only strategically relevant problems of corporate products and services were focused as potentials for improvement. Today, the complaints of customers, evaluations of managers, and notes from employees serve to identify improvement potentials. Relevant and suitable potentials are described as business cases for Six Sigma projects. These projects are then supervised by managers in the role of a project Sponsor, and implemented by Green or Black Belts, together with the necessary subject matter experts according to the DMAIC method.
2. The DMAIC is a methodical approach for the implementation of improvement projects. At its core, the DMAIC cycle is a generic, hypothesis-based, scientific approach to problem-solving. It is based on an open box of chronologically linked rational-logical and statistical tools to uncover problem–cause relationships.

[1] Video 3 (https://www.youtube.com/channel/UCShsy50l5dFsoSXU__hBKSA)

© Springer Nature Switzerland AG 2019

R. Hutwelker, *Six Sigma Green Belt Certification Project*, Management for Professionals, https://doi.org/10.1007/978-3-030-31915-1_2

1. Enterprise Improvement Mechanism: increase customer satisfaction and reduce costs

- Problems of corporate products & services offer potentials for the improvement of processes
- These improvement potentials are identified by customer, by the management and by employees
- Potentials are defined as Six Sigma projects and then supervised by *Sponsors* from management
- Projects are implemented by Green-/ Black-Belts, according to the method and tools of the DMAIC cycle

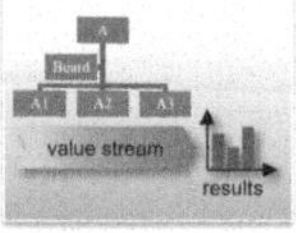

2. Methodical Approach: chronologically linked tools for the implementation of the project

- The DMAIC-Cycle is a problem-solving approach in five phases for the implementation of projects
- The approach is based on an open box of chronologically linked rational-logical and statistical tools
- The tools serve to define problems, analyze causes, develop solutions and control their success
- Additional tools serve to prepare the business case and to manage the project

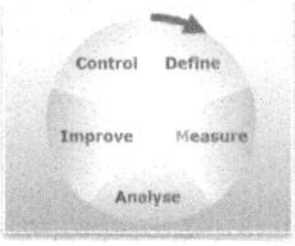

3. Statistical Basis: process capability, process control, problem-cause relations, improvements

- The sigma level is a performance measure of the process capability, i.e. the excellence of the outputs
- At a level of 6 sigma (6σ), only 3.4 errors are expected in 1 million outputs (Motorola's short-term-sigma)
- Control charts indicate the variation of the process performance, i.e. variation of the output over time
- Statistical tests identify the main problems, the relationships to their causes and degree of improvement

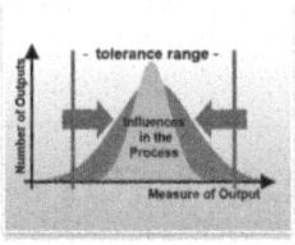

Fig. 2.1 Six Sigma is a concept with several meanings—it links practice with theory and statistics to analyze and realize potentials and to verify improvements

Overall, the chain of tools is designed to define problems, analyze causes, develop solutions, and control their success. Additional tools help to formulate the business case, to organize the project environment, and to manage the project.

3. The Sigma Level is a performance measure of the process capability, i.e., the excellence of the output. At a performance level of 6 sigma only 3.4 errors are expected in 1 million outputs (Motorola's process-sigma). Control-Charts indicate the variation of the process performance, i.e., variation of the output over the time. Statistical tests evaluate formulated hypothesis about the size of problems and the relationships to their causes. They also provide a statistically verified basis to evaluate the success and the benefits achieved.

Six Sigma is a concept with several meanings. It links practice with a method to rationally and statistically analyze and realize improvement potentials.

2.2 Company-Wide Improvement Program[2]

If a company's processes are to be systematically improved with Six Sigma, then an appropriate program must-be implemented. Approximately 1% of employees are to be trained as Black Belts and 10% as Green Belts. Of course, these Belts should not only implement their certification project. At least one project per year would be desirable. And the organization of Six Sigma as a company-wide program can ensure this (Fig. 2.2).

[2]Video 4 (https://www.youtube.com/channel/UCShsy50l5dFsoSXU__hBKSA)

Success factors for Six Sigma Programs:
- Involvement of the management via policy deployment and target agreement
- Specific career options for successful Green and Black Belts

Number of projects implemented per Green Belt:
- In successful programs: 1.6 projects per year (Mean)
- In terminated programs: 1.2 projects in the lifespan of the Six Sigma program (Mean)

Unsuccessful Six Sigma Programs:
- Termination after 5.7 years (Mean)
- Reasons for termination:
  - new management appointed, e.g. after reengineering of the company or mergers
  - no more suitable improvement topics identified, conviction that nothing more can be improved

Fig. 2.2 Enterprise Improvement Mechanism: The quality of a Six Sigma program is determined by the management, career options for Belts, and the continuous identification of suitable project topics

Our study on Six Sigma programs in Germany has shown—Leyendecker, B. Schindewolf, S., Hutwelker, R, & Weigel, H. (2011)—that the involvement of management through policy deployment and target agreements is an important success factor [best practice]. Another success factor is the prospect of improved career opportunities for Green and Black Belts [best practice].

With successful programs, a Green Belt carries out an average of 1.6 projects per year and the efficiency of implementation increases with experience. For less successful programs, each Green Belt only carries out a total of 1.2 projects during the life span of the Six Sigma program.

If Six Sigma programs are terminated, then on average after 5–6 years. One reason for this is a new management that wants to set new accents. Another reason is that no more suitable project topics are found. This is usually the case if the company focuses exclusively on strategically relevant improvement issues [bad practice]. However, successful programs are characterized by the fact that all relevant stakeholders are involved in identifying topics.

One source for improvement topics is the management and its knowledge about: target achievement, deviations in further key figures they are responsible for, and its observations about the effectiveness and efficiency of their processes. Another source are customers, their ratings and complaints, their purchases and cancellations. Employees have a direct view of the processes. They must clarify the inputs they receive; they experience the negative influences in the process, and they can directly detect deficient outputs (Fig. 2.3).

It is also possible to identify improvement topics by specific activities. They can result from filtering the suggestions of the suggestion system. And topics can often be found in Value-Stream-Analysis workshops and sometimes also in 5S workshops.

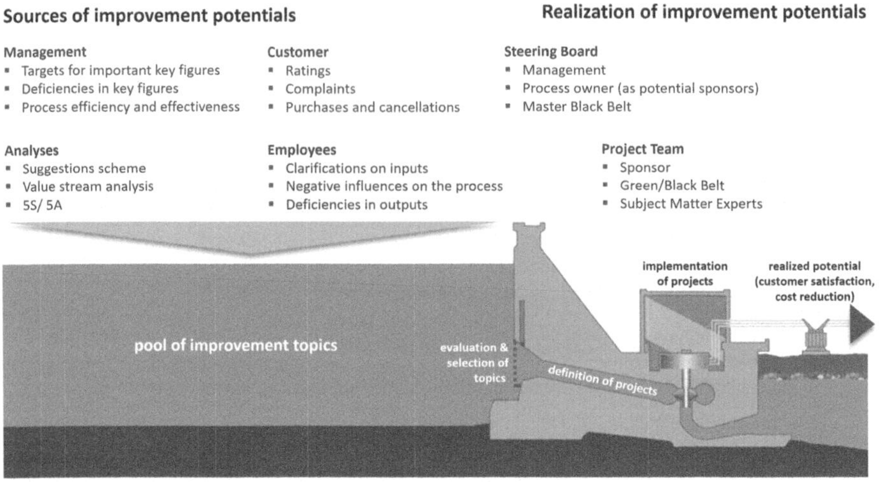

Fig. 2.3 The sustainability of a Six Sigma corporate program is ensured by the continuous identification of improvement topics and their implementation in projects (Graphic of hydropower plant adapted from: https://commons.wikimedia.org/wiki/File:Hydroelectric_dam_german.png)

All identified topics are collected in a pool. The topics of the pool are regularly evaluated according to their potential and their suitability for Six Sigma. This evaluation is best performed by the members of a Six Sigma Board, consisting of management representatives, process owners as potential project Sponsors and Six Sigma experts, like Master Black Belts. I know a company where this selection process is transmitted live to the employees on their intranet to make these decisions of the Six Sigma Board Meeting transparent. If an employee's project topic is selected there, he or she automatically becomes a Green Belt candidate. This is the reward for the employee who walks through the company with open eyes [best practice].

After selection, Six Sigma experts define the selected topics as business cases for Six Sigma projects. These projects are then implemented by project teams, consisting of a Sponsor, a Green or Black Belt, and experts from the process concerned. Through this mechanism, the significant potentials of a company can be realized step by step, project by project.

However, you do not have to involve the entire organization right from the start. It is enough to find a suitable topic at the beginning and convince a responsible manager to approve and support your project in the role of a Six Sigma Sponsor. That would be a good start which we want to recommend and support in this Green Belt course. More on this is in the next chapters.

Questions (SC= single choice; MC= multiple choice)

Video 4/ Ch. 2.2/ #1 (SC)	Please assign the roles to their explicit/implicit tasks in a Six Sigma program.				
		Identify project topics	Select projects	Support projects	Lead projects
1	Black Belts				
2	Customer				
3	Employees				
4	Green Belts				
5	Manager				
6	Master Black Belts				
7	Sponsors				
8	Subject Matter Experts				
9	Yellow Belts				
9	points				

2.3 Practice and Application: Cookie du Chef[3]

Wanda will show you how to make cookies. She demonstrates the problems of the outputs and their causes in the process (Fig. 2.4).

Fig. 2.4 *Wanda*, the three-year-old chef of our bakery*: Cookie du Chef* will demonstrate the business case for our example project

[3]Video 5 (https://www.youtube.com/channel/UCShsy50l5dFsoSXU__hBKSA)

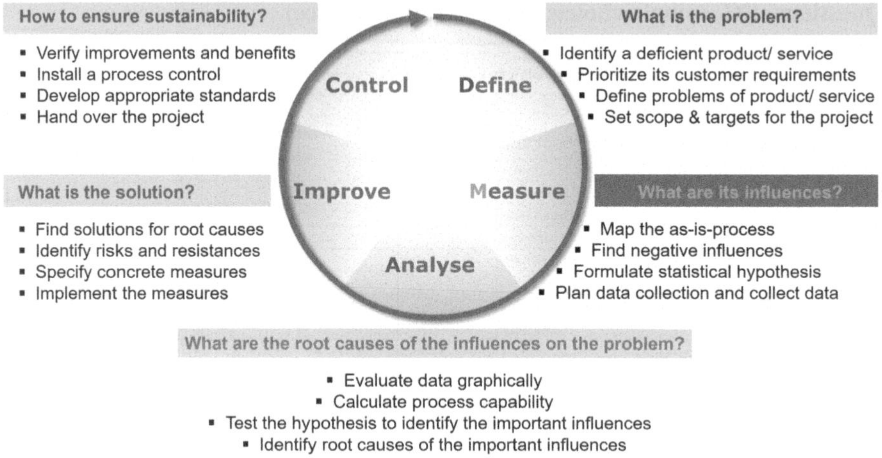

Fig. 2.5 DMAIC phases with chronologically linked rational and statistical tools for implementing the project

2.4 Six Sigma Method

2.4.1 Phases of the DMAIC Cycle[4]

Each Six Sigma project follows the phases of the DMAIC cycle, like in a screenplay. The sequence of these phases determines the dramaturgy and the tools in each phase determine the scenes of the plot (Fig. 2.5).

Six Sigma projects are processed along the phases of the DMAIC cycle (DEFINE–MEASURE–ANALYSE–IMPROVE–CONTROL). And each phase covers a set of chronologically interlinked rational and statistical tools.

In the **DEFINE** phase we answer the question: What is the problem? First, a product/service that deviates from the requirements must be identified as a potential for improvement. Customer requirements are then prioritized as a basis for defining the primary problems of the product/service. The specified customer requirements and the size of the problems help to define the business case for the project. Finally, the scope and objectives of the project are defined, and a project team is formed with subject matter experts as supporters for the project.

After defining the problems, we look for their triggering causes in the **MEASURE** phase. To identify these negative influences on the problem, the as-is process is mapped and inspected first. The next step is to formulate hypotheses that reflect the relationship between the triggering influences and the problem. Then a plan can be created to collect data on the problems and their influences and finally the specified data are collected.

[4]Video 6 (https://www.youtube.com/channel/UCShsy50l5dFsoSXU__hBKSA)

In the **ANALYSE** phase we answer the question: What are the root causes of the influences on the problem? At first the measured data are evaluated graphically. Next the process capability is calculated and—most important—the hypotheses to identify the significant influences are tested. Either on a statistical or on a rational base the root causes of these influences are finally identified. They are the key to solving the problems.

In the **IMPROVE** phase we answer the question: What is the solution? At first, solutions are developed to eliminate, adjust, or circumvent the identified root causes. Risks and resistances against these solutions are then identified and the solutions are adapted to put them into practice. The adapted solutions are then specified as concrete measures with responsibilities and deadlines. The Sponsor then decides on their implementation and the measures with a positive decision are finally implemented.

In the **CONTROL** phase, we answer the question: How can we ensure the sustainability of our improvements? First, the degree of improvement is calculated on a statistical basis and the achieved financial benefit is calculated. Then a Process-Management-Plan is installed, with Control-Charts to monitor the process and with countermeasures for deviations that occur. New standards for the process are developed and communicated. Finally, the project is handed over to the process owner.

Each Six Sigma project is based on a process-problem-solving model. Here the framework of the plot is structured and the relationship between the actors is defined.

Questions (SC= single choice; MC= multiple choice)

Video 6/ Ch. 2.4.1/ #1 (SC)	**Please assign the tasks of a Six Sigma project to it's DMAIC phases.**					
		DEFINE	MEASURE	ANALSYE	IMPROVE	CONTROL
1	Identify project topic					
2	Calculate benefits of the project					
3	Collect data					
4	Determine the problems (Y)					
5	Develop a process-management-plan					
6	Develop solutions					
7	Formulate hypotheses					
8	Identify the influences (x)					
9	Identify the risks of measures					
10	Identify the root causes (x')					
11	Identify requirements on the product/ service					
12	Implement concrete measures					
13	Listen to the Voice of Customer					
14	Test hypotheses					
14	points					

2.4.2 Process-Problem-Solving Model[5]

Process Model

Our business case here is about baking cookies. And I am sure you have at least seen the process of how cookies are baked before. Let us change from this reality section to the method. Baking cookies involves several things: sugar, a measuring jug, operation manuals, a timer, rolling the dough, an order, rules and standards, an oven, a whisk, a rolling pin, eggs, chocolate, kneading the dough, a kitchen scale, our chef and her manager, energy, a recipe and the cookie. All pictures show elements of baking. But what is their role and how do they relate to each other in this process? (Fig. 2.6)

The **Output** is the most important element of a process. This is what the customer receives, pays for, experiences, and evaluates. The outputs of our bakery are of course the *cookies*.

To create one or more cookies the corresponding **Inputs** are needed—the *order* and the *ingredients* for cookies in our case.

Between input and output there is a certain **Activity** which transforms the input into the output, e.g., *kneading the dough* or *rolling the dough*. You can identify the activity by asking: "What is done?" or more general: "What happens?" Please always describe the activity with a verb in the present progressive tense to express that something is really done here [best practice].

There must be of course someone and/or something that executes the activity. Employees, machines, tools, and energy might be these **Resources** of the process. Here you will find our *Chef, the timer, the whisk, energy,* and the other resources, through which and with which the activity is carried out (Fig. 2.7).

Additionally, a **Method** is needed to guide the resources for a certain activity. In our case the used methods are: the *recipe*, the *operating instructions* for our equipment, and some *rules* and *standards* that apply in the kitchen.

[5] Video 7 (https://www.youtube.com/channel/UCShsy50l5dFsoSXU__hBKSA)

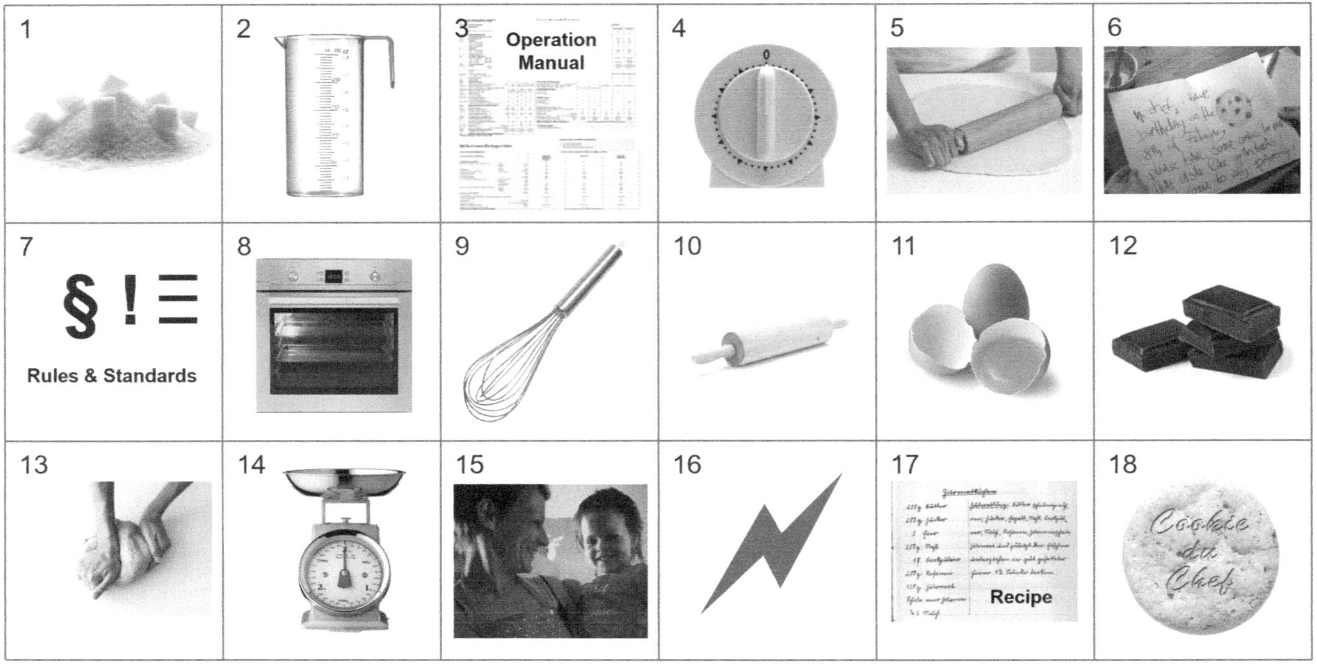

Fig. 2.6 Several *ingredients* are involved in the process of baking cookies. What is their role and how do they relate in this process?

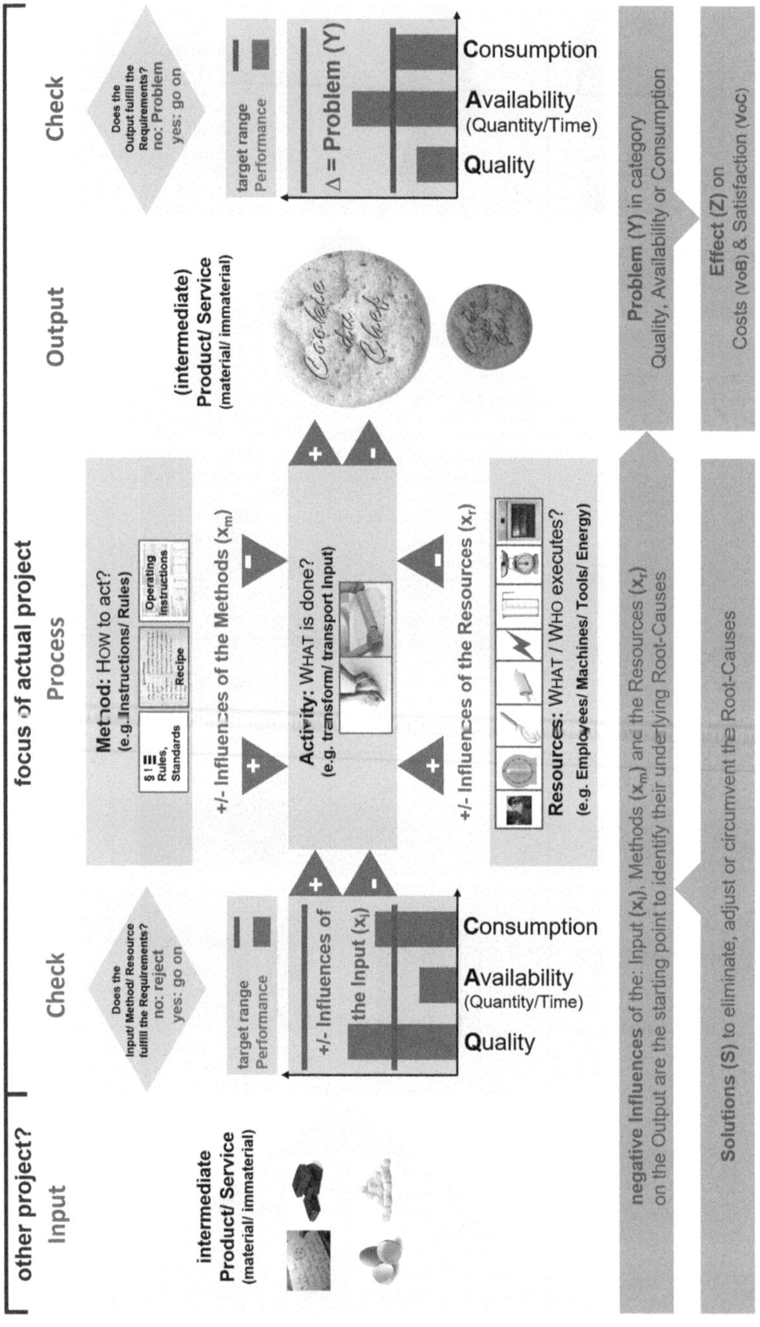

Fig. 2.7 Problems are deviations of the output in important attributes. Their triggers are in the input, in the method, and the resources that direct the activity

These are the necessary components of every process: input, activity, method, resources, and output.

Details: There Are Many Possibilities to Act: One of Them Is Best Practice

In many companies, however, the activities are only partly—or not at all—guided by specified methods. The management often relies on the common sense and the intelligence of their employees and often only documents the activities of processes [bad practice].

However, the differences in the intelligence and common sense of different employees lead to a variable execution of the activity and thus to variations in the output. That does not have to be significant. But in any case, different processes produce different results, even if the performance indicators of the output do not reflect this variation. Think, for example, of the extreme preheating of the oven before baking in the video (Sect. 2.3).

Please note that the methods bring the know-how into your processes and preferably the best practice methods. Just think of the *checklists for pilots* [best practice]. As far as possible, methods should thus be standardized and continuously developed to improve the activities and the resulting outputs—except of course if creative activities are necessary.

Additionally, some resources like energy and operating supplies are often considered as inputs and thus, the borders between inputs and resources become blurred.

If you, however, accept that the activity transforms or transfers the input into an output, then you can easily differentiate between input and resources. Outputs are the transformed or transferred inputs.

This concludes the description of the process part of the model. Next, the problems and causes are anchored in it.

Basics of Problem-Solving

All customers have individual **Requirements** for the outputs. You, e.g., will probably review and evaluate the taste, ingredients, weight, size, color, and texture of cookies you bought. These requirements are aimed at the: **Quality** attributes of the cookies. However, the customer may also need a certain number of cookies at a certain time and place. These are the requirements for the: **Availability** of cookies. At least the customer may only spend a certain amount of money for the cookies. This attribute belongs to the category of: **Consumption**.

The company also has output requirements. Our bakery wants to produce cookies with low input and resource consumption and therefore does not want to waste, e.g., any ingredients or energy. The customer and the company thus evaluate how well the requirements are met. Both evaluate the outputs based on their important attributes and our cookies must meet these requirements within a specified target range. In other words: Customers and the management of a company compare the actually achieved values of important cookie attributes with a target range for these attributes.

Within the bakery we will also check important attributes of all inputs, e.g., the expiry date of the eggs or the taste of the chocolate. Chocolate and eggs are outputs of the suppliers that our bakery in the role of a customer needs as inputs. According to a personal survey, employees use between 10% and 30% of their working time to deal with inadequate inputs to carry out their processes properly [bad practice].

In this model we first focus on *our* customers who receive *our* output and *our* company that is responsible for this output. If we, e.g., produce a burnt cookie, then this is a **Problem** because it indicates a deviation between the requirement on the cookie's taste and the actual manifestation of the taste. Generally speaking: If our customer or our company is not satisfied with the cookies because their quality, availability, and consumption are outside the tolerance corridor, then this is a problem. A Six Sigma project always starts with a problem. And the problem in Six Sigma is like a nail in the wall to which the whole project hangs.

What are the **Causes** of the problems? The activity generates the output. But the activity is completely determined by the resources and methods. If the activity in a process step is unsuitable for achieving the desired output, the negative Influence either results from a resource that deviates from a defined method or the method itself is unsuitable, such as an incorrect recipe. If no method at all is defined, e.g., how the size of the cookies is determined, then the employee follows his individual method as to how the activity is to be executed. This made the cookies too big for the can.

The problem of the burnt cookie can, e.g., be caused by a wrong recipe by our Chef who does not adhere to the recipe by a defect timer or oven. The negative influence, therefore, lies in an unsuitable resource, in an unsuitable method, or in a deviation from a suitable method. A rotten egg we receive triggers another problem in our output. Here the negative influence lies in the unsuitable input.

Input, method, and resource are, therefore, the sources of negative influences on the output. The root causes of these triggering causes, however, can of course lie outside the focused process. In our project we must solve the problems of our outputs. Typically, we cannot solve the problem of deviant inputs like the rotten eggs we received as input. This is the task of our supplier. However, we can in the future circumvent this negative influence by drawing the supplier's attention to his problem or by buying the eggs from another supplier.

This differentiation draws the limits of our project. If our process leads to a problem in the output, then the scope of our project lies in the process between the inputs that we cannot influence up to the output with the identified problem.

Every problem has **Effects**. It can determine the customer's satisfaction with the output. To know this effect, you must listen to the Voice of the Customer (VoC). Problems also have a negative impact on the costs of an output. To get to know this effect, you must listen to the Voice of Business (VoB).

What can be done to solve the problems and thus avoid their effects? The answer is simple. If you know the root causes of the negative influences, then develop

Solutions to eliminate, adjust, or circumvent the root causes. If the real root causes are known, then it is often quite easy to handle them just by turning the root causes upside down.

This process-problem-solving approach is the basis of our Six Sigma project. We start with the problems and they are called **Y**s. Their negative influences, from the trigger to the underlying root causes, are called **x**s. This notation results in the basic function for statistical analyses: **Y**= f (**x**).

Each Six Sigma project is based on definitions and principles. Here the characters of the plot and their dialogues are detailed.

Questions (SC= single choice; MC= multiple choice)

Video 7/ Ch. 2.4.2/ #1 (SC)	In which part of a process can influences/causes (x) occur and in which do problems (Y) occur?					
		Input	Method	Activity	Ressources	Output
1	Influences (x)/ Causes (x')					
2	Problems (Y)					
2	points					

Video 7/ Ch. 2.4.2/ #2 (SC)	To which part of a process do these components of our bakery belong?					
		Input	Method	Activity	Ressources	Output
1	Vanilla					
2	Energy					
3	Write shopping list					
4	Chef					
5	Cookie					
6	Oven manual					
7	Enter the order					
8	Order					
9	Cookie delivery note					
10	Hygiene regulations					
10	points					

Video 7/ Ch. 2.4.2/ #3 (MC)	What types of problems can an output have?					
		Quality	Costs	Satisfaction	Availability	Consumption
1	An Output can have a problem in terms of:					
1	points					

2.4.3 Definitions and Principles of Problem-Solving[6]

Searching for suitable topics I always received a lot of information about the organization: lack of communication and cooperation, deficient interfaces, unreliable (internal) suppliers, and a desire to reduce costs. But that did not really help me and it is difficult to filter suitable topics out of this information. That is why I turned the tables at one point: I will ask the questions politely, of course.

The following details and examples can help you to identify a suitable project topic.

[6]Video 8 (https://www.youtube.com/channel/UCShsy50l5dFsoSXU__hBKSA)

Six Sigma projects start with a **Problem**. And problems are attached to outputs. Thus, you need to identify a critical output, which can be a final or an intermediate product or service.

1. Outputs are real or virtual objects. They can always be indicated by a just noun or a compound noun, like: Shopping list, ingredient, dough-ball, dough-sheet, cookie-blank, cookie.
2. Outputs are countable as they result from a repetitive process. In a flow production with continuous output, the product can be a container. In the service sector: Contracts, invoices, data entries, answers, decisions, deliveries, and even patients are countable outputs.
3. If you complete the sentence: *The output* ***is*** ... then you define the object itself (e.g., cookie **is** a foodstuff). If you complete the sentence: *The output* ***has*** ... then you describe its attributes (e.g., cookie **has** a taste, weight, calories, storage location).
4. Neither a cookie or a shopping list itself can be a problem, nor other products and services. These outputs, however, can have problems if their attributes deviate from requirements (Figs. 2.8 and 2.9).

   Imagine that all outputs carry a label with all its characteristic **Attributes**, like a label on food packages, but more detailed, which indicate ingredients, date of manufacture, expiry date, weight, size, color, baking time, oven temperature, consumed energy, working hours required, and so on. Every product and every service could have such a label with all its attributes—this contribution in my video, as well as all your outputs. What attributes would appear on the imaginary label of one of your outputs? This is not always as easy as with cookies. But it is

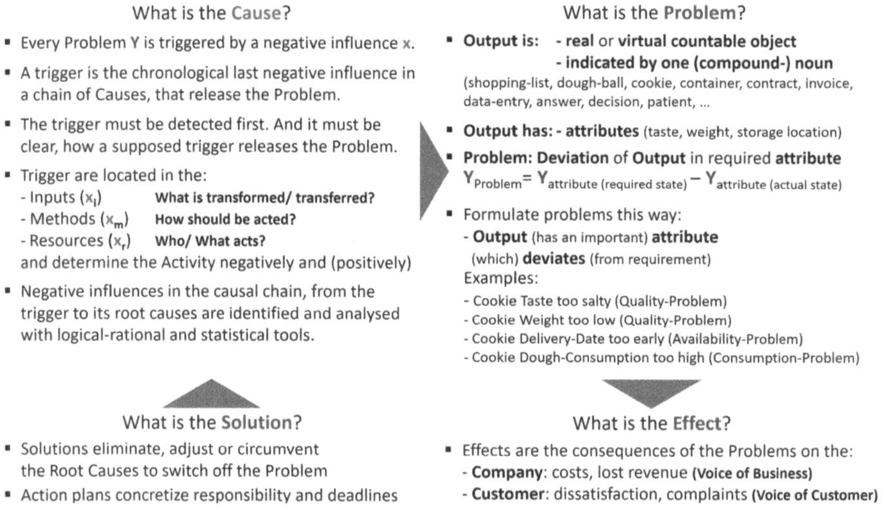

Fig. 2.8 Every Six Sigma project starts with a problem. The first challenge is to identify the problem in the available information on the issue

Attributes each cookie has:

- Ingredients
- Date of manufacture
- Expiry date
- Weight
- Size
- Color
- Baking time
- Oven temperature
- Consumed energy
- Working hours required
- Consumed dough
- Taste
- Cycle time
- Storage location
- Storage period
- Serial number
- Calories
- Date of sale
- Price
- ...

Fig. 2.9 List of cookie attributes. Similar labels with attributes of quality, availability, and consumption could be assigned to every tangible product and intangible service. The requirements of the customer and the company determine the importance of these attributes (CtQs)

important and useful to think in terms of attributes. Outputs cannot be measured, but their attributes can.

You will remember that a problem is defined by the deviation of an output in an important attribute, and more specific:

5. A problem is the difference between the required state of the attribute and the actual state of the attribute. If this difference exceeds a generally accepted or even specified limit, then the output has a problem. Do you know of problems of outputs in your work environment?
6. Every problem can be formulated within this framework:
   - **Output** (has an) **Attribute** (which) **deviates from requirements**.

     For example, Cookie Taste too Salty—this is a quality problem, Cookie Delivery-Date too Late—this is an availability problem, or Cookie Dough-Consumption too High—this is a consumption problem. Please try to express each of your identified problems in this format.

What is the **Effect**? Effects are the consequences of the problem on the company and the customer. Typical effects on the company are costs and lost revenues. They determine the business case and need to be specified as Voice of Business (VoB). Typical effects on the customer are dissatisfaction and complaints. They need to be specified as Voice of Customer (VoC).

What is the **Cause**? Walter A. Shewhart (1939) discovered that variations in the process lead to variations in the output. Thus, we must uncover the relationship between the problem and its root cause. How can we be sure to detect these root causes?

The incident analysis offers a suitable procedure for this purpose. Each problem must be triggered by a last negative influence in a chronological causal chain. To find the root causes, we must analyze the entire causal chain from the trigger that releases the problem to an underlying cause that we can control. Thus, it must always be clear for you, even if you are not a subject matter expert, whether and how it can be that a supposed trigger, like, e.g., a rotten egg from the input, or a wrong recipe, or a defect timer releases the problem, e.g., Cookie taste too salty. We will later come back to this topic and analyze the causal chain from the trigger to its root causes with logical-rational and statistical tools (Fig. 2.10).

What is the **Solution**? Solutions eliminate, adjust, or circumvent the root causes to switch off the problem. Thus, root causes are the anchors for solutions. Solutions can be distinguished from concrete, practicable measures by defined responsibilities and deadlines for their implementation.

Now the problem-solving approach within the process model becomes apparent: The root cause and chronologically following intermediate causes trigger the problem and the problem has an effect.

Our entire daily routine can be divided into inputs, activities, outputs, and problem-solving, even if we are not always aware of it. We start and finish the day

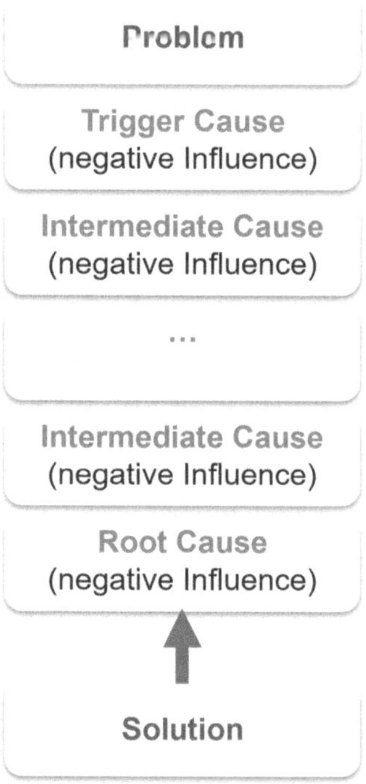

Fig. 2.10 Hierarchical causal chain, from the problem of the output to the triggering cause in the activities or input down to the root causes, which serve as interfaces for solutions

with the problem: Tooth Plaque layer too Thick. You know the effect of this problem (caries) and its solution (brushing teeth). You know the necessary input, your activity, the resources, and the method for this service process. You probably also know the best practice method for tooth brushing to achieve: Teeth Plaque layer acceptable. How about listing additional outputs—especially from your work environment—and anchoring them in the process-problem-solving model.

Questions (SC= single choice; MC= multiple choice)

Video 8/ Ch. 2.4.3/ #1 (SC)	Please classify these observations from our bakery into the categories of the process-problem-solving model.				
		Influence (x)	Problem (Y)	Effect (Z)	Solution (S)
1	Cookies surface broken				
2	Vanilla not available				
3	Cookie diameter too large				
4	Customer complaints				
5	Cookie invoice wrong				
6	Adhere to recipe specifications				
7	Kitchen scale incorrectly calibrated				
8	Forbid nibbling on the dough				
9	Egg-yolk/ egg-white separation laborious				
10	Decline in sales of cookies				
11	Change chocolate supplier				
12	Energy costs too high				
12	points				

2.5 Statistical Basis of Six Sigma[7]

2.5.1 Basics of Process Capability

You know that the term Six Sigma also indicates a specific level of process capability: Sigma level= 6. At this level only 3.4 errors are expected in 1 million outputs (= 99.99966% yield), according to Motorola's Long-Term Sigma (see below). Process capability generally reflects the performance of a process to meet the requirements for an important attribute of the output. What is behind this evaluation?

Weight is an attribute of cookies and we promise that our cookies have an average weight of 10 g. Of course, we cannot always bake exactly the same cookies. Therefore, we additionally promise to keep the limits between 7 g (Lower Specification Limit) and 13 g (Upper Specification Limit). To check how good we fulfill our promise, we scaled each cookie and assigned it to one of different weight categories: 3–5, 5–7, 7–9 g and so on by drawing a bar. Stacking the bars in the corresponding weight category results in the displayed frequency distribution in the Histogram. And this sample fits well with the bell-shaped model of the normal distribution of Carl Friedrich Gauss (Fig. 2.11).

[7]Video 9 (https://www.youtube.com/channel/UCShsy50l5dFsoSXU__hBKSA)

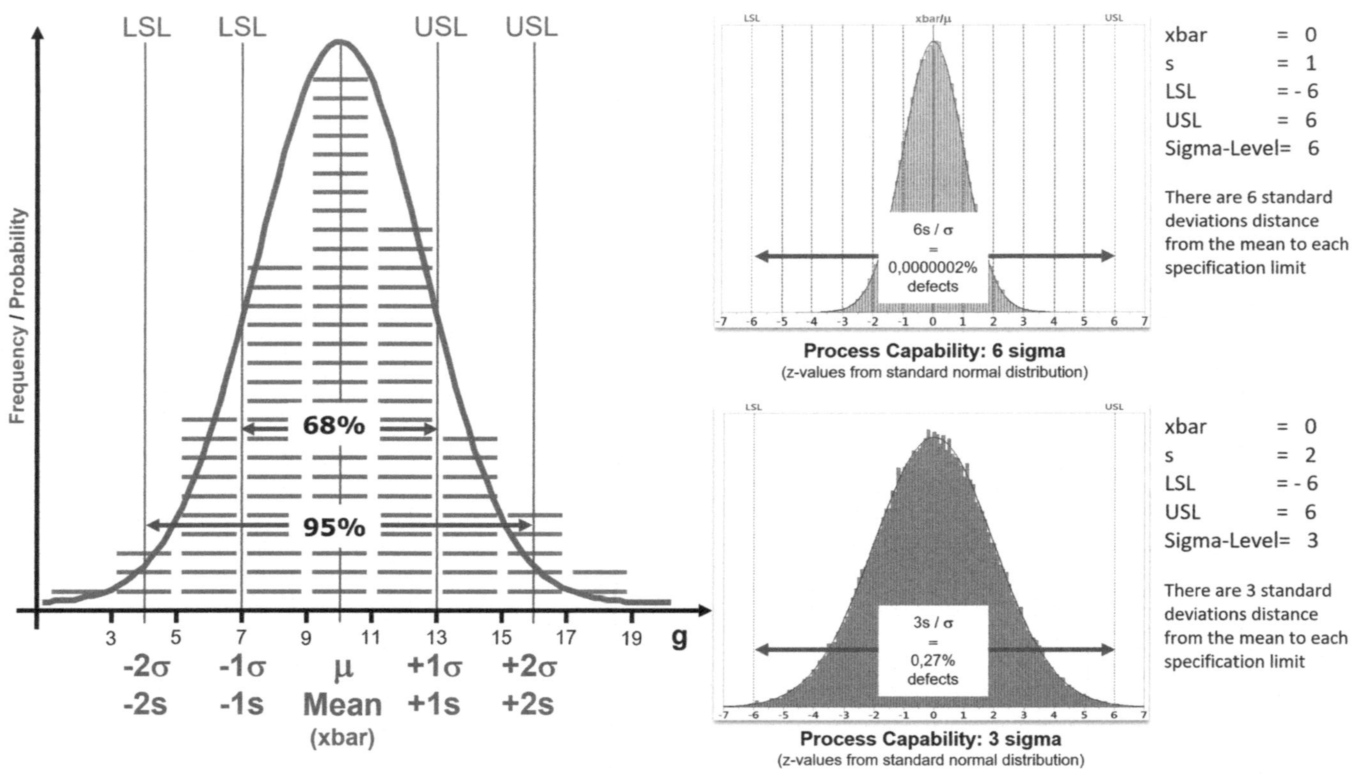

Fig. 2.11 Output requirements are specified by a tolerance range for each important attribute (LSL–USL). The more outputs are within the Lower (LSL) and Upper Specification Limits (USL), the more capable the process is

You can see that the average weight of cookies in this sample is 10 g (xbar= 10 g), and that their weight is scattered around the mean with a standard deviation of 3 g ($s = 3$ g). You probably know that the area of the distribution between minus one to plus one standard deviation around the mean covers 68.27% of the cookies.

Accidentally in our example, the Lower Specification Limit falls at −1 s (10 g – 3 g= 7 g = LSL) and the Upper Specification Limit at +1 s (10 g+ 3 g= 13 g = USL). Thus, our actual process has a process capability of 1 sigma, which is equivalent to a yield of about 68%.

Mean and standard deviation and the specification limits are thus necessary to calculate the process capability.

To understand the concept of the mean and the standard deviation, it is useful to know both functions by heart. And the mean can easily be identified at the center of the symmetric normal distribution. Less known is that one standard deviation (s= 1) also results from the perpendicular through the inflection point of the distribution curve. The interval between the mean and this vertical line on the x-axis corresponds to one standard deviation (s= 1). This allows you to quickly estimate one and more standard deviations in a Histogram.

You probably also remember that the area between minus two (−2 s) and plus two standard deviations (+2 s) covers about 95% of the cases and 99.73% of our cookies fall between −3s and +3s. If now, accidentally, the lower (LSL) and upper specification limits (USL) for our cookies fall on −3 and +3 standard deviations, then our process has a capability of 3 sigma, which is equivalent to a yield of 99.73%. The more outputs are between and the fewer are outside the specification limits, the more capable the process is. The capability decreases with the number of cookies falling outside the specification limits—and each of these outsiders has a problem.

This principle is further illustrated by the simulated 6 sigma process (short term). There is an interval width of 6 s from the mean value to the specification limits. And as you can see, there is still a lot of space from the extreme values of the distribution up to these respective specification limits. Now compare these relationships with the simulated 3 sigma process. The extreme values of this distribution are already outside the specification limits and it can easily be evaluated that the 3 sigma process (short term) is less capable than the 6 sigma process.

But why do we call this capability 3 sigma and not 3 standard deviations? If we calculate parameters for a certain sample or plot, e.g., a concrete frequency distribution (grey), then we use Latin letters (xbar; s). Greek letters (μ; σ) and the corresponding calculations (confidence intervals) are used in statistics if we want to infer from the concrete sample we took to the assumed population of possible outputs (red). This is what the customer can expect in the future (red), derived from the collected sample (grey).

2.5.2 Influences on Process Capability

Details: Shooting at a Wall and Drawing a Target Around the Hole or Vice Versa?

There are two ways to increase process capability:

1. If we convince our customer to widen the area of tolerated outputs, i.e., shift the specification limits away from the mean, then the capability of our process rises without changing the process.

   It is just because we got a broader garage without improving our capability to drive the car. This proposal is not entirely serious but may be necessary in cases where different departments subtract individual fear buffers from the specification limits originally required by the customer.

   [Practice: In a few cases it makes sense to question the reasonableness of given specification limits. Does the customer really have such strict requirements?]
2. Typically, you must improve the process by centering your outputs and reducing their variability, so that more outputs fall between the specification limits. In this case you always also get into a very narrow garage without dents. [Practice: Typically, specification limits are immovable, like the side walls of a garage.]

Letus apply this idea to our bakery. Imagine we baked cookies on four days, Monday, Tuesday, Wednesday, and Thursday. We have measured the weight and diameter of the cookies and mapped the weight on the horizontal and the diameter on the vertical axis. We have also drawn a target and promise our customer that weight and diameter will be within the second ring of the target, representing our specification limits (Fig. 2.12).

What differences do you notice between the results of the four days?

The results on Monday and Wednesday differ from the results on Tuesday and Thursday (columns) in the degree of their scattering. And the results on Monday and Tuesday differ from the results on Wednesday and Thursday (rows) in the degree of their position centering.

If we now drop the weight data on the x-axis and stack them, as shown in the last slide, then again the idealized distributions of the weight result for each of the four days. If we now project the specification limits onto the weight-axis, then we can already estimate the portion that falls within and outside the specification limits on the respective days. On Wednesday, e.g., we achieved a result of about 1 sigma (perpendicular at inflection points) and Thursday was our best day.

Today is Friday: You are the Six Sigma expert and all responsible process owners are sitting around you. Everyone agrees that Thursday's results should be repeated in future. Unfortunately, we lost Thursday's records of process and machine settings.

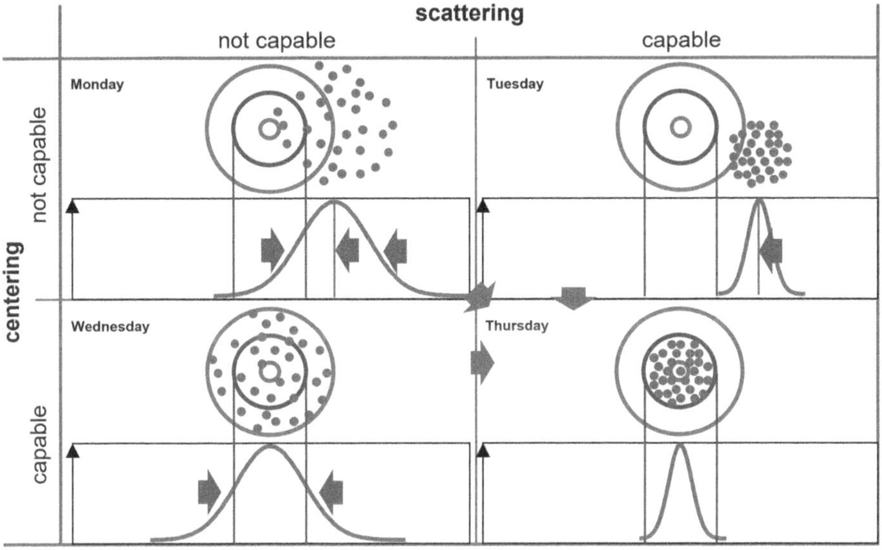

Fig. 2.12 Six Sigma's goal is to align performance indicators of products and services to the center and to minimize their variability, according to customer requirements

But we have the records for the other three days. Which day offers the best starting point to achieve the good result of Thursday again?

You would probably not start with Monday's settings. But what about Wednesday? We have a large spread, but we produce about 68% suitable cookies right away. If we start with Tuesday's settings, then we only produce scrap, but we have already achieved a very small spread. How would you decide? What would be the best starting point?

If we have no idea which influences were responsible for the position shift (Tuesday) or the scattering (Wednesday), then we would have to assume how shifts and scattering are typically caused. If we are lucky then the position shift is caused by exactly one systematic influence [Practice: Shifts are typically caused by one/a few systematic influences]. Once we have identified this, we can correct the process quickly and with little effort. Systematic influences in the process, e.g., shifts, are indicated in Control-Charts.

Scattering, however, is almost always caused by an interaction of a multitude of influences. This is the basis for the normal distribution in many application areas. Think of the many and complex influences that determine, e.g., the human body size. And it is not always easy to identify and control these influences. So, if we could choose a day, we would choose Tuesday.

But we cannot do that in practice. Typically, we will have to deal with scattering and position shifts. This is the typical and challenging task in a Six Sigma project: centering the position and reducing the scatter. Nothing more, but nothing less.

2.5.3 Examples of Performance Levels in Everyday Life

To give you an impression of the relative proportions of percentages and Sigma Levels, I researched the defect rates for a few everyday examples and converted them into Sigma Levels. However, these are not statements about process capability, as no specification limits are considered. It is only a conversion of percentage values into z-values of the standard normal distribution. And strictly speaking, this is not allowed either, because defect rates are not and cannot be normally distributed. More about this later in the measure phase. Nevertheless, I would like to briefly explain the difference between short-term and long-term sigma using these examples.

Each example (see Table 2.1) shows what a yield of 99% would mean. I also documented from different sources what actually happens in reality. Here you can compare the actual yields in percent and the corresponding sigma values for long-term and short-term observations, respectively.

As you can see, a 99% yield in air traffic would probably frighten us so much that no one would fly voluntarily. In fact, however, safety in air traffic is 99.9999894%, which corresponds to a long-term Sigma Level of 5.19. By the way, a high sigma value is easier to communicate than a corresponding percentage with many decimal places. Safety at the workplace and in hospitals is also very high, but already significantly lower.

The reference example uses a typical paperback book with 300 words per page and 300 pages per book to show the standard example for short-term 6 sigma, with 3.4 errors per 1 million words, i.e., 1 error in 3.27 books. This performance level would have to be adjusted to 4.5 sigma if the long-term performance is to be estimated.

What does the differentiation between short- and long-term Sigma Levels mean? Observations of process performance over longer time intervals show that the mean

Table 2.1 Examples for Sigma Levels and corresponding percentages in everyday life

examples from everyday life	99% yield *would* mean:	in reality happend:	Yield (%) short-term long-term	Sigma Level (short-term)	Sigma Level (long-term)
Air traffic safety	11 crashes of aircraft daily arriving and departing Munich Airport	37.6 million flights with 4 fatal accidents (2015, worldwide)	99,9999894%	(6,69)	5,19
U-train victims of violence	1 victim/ 100 passengers	139 victims/ 349 million passengers (Germany)	99,9999602%	(6,44)	4,94
Reference: Errors in typical paperback (300 words/page; 300 pages/book)	**3 errors per page; 900 errors per book**	**6 sigma performance corresponds to: 1 error per 3,27 books**	**99,9996600%**	**6 [1)]**	**4,5 [2)]**
Work safety	10,000 industrial accidents/ 1 million working hours	14,76 industrial accidents/ 1 million working hours (Germany)	99,9985240%	(5,68)	4,18
Counterfeit money	10,000 counterfeit notes/ 1 million bank notes	50 counterfeit notes/ 1 million bank notes (Europe)	99,9950000%	(5,39)	3,89
Error in hospital treatment resulting in death	190,000 errors resulting death/ 19 million hospital treatments	19,000 errors resulting death/ 19 million hospital treatments (Germany)	99,9000000%	(4,59)	3,09
Driving licences revoked for alcohol	62000 revoked licenses / 6.2 million visitors	158 revoked licenses/ 6.2 million visitors on Oktoberfest 2017	99,9974516%	4,05	(2,55)
Executed death penalty on innocent people	1 innocent/ 100 executed	4 innocent/ 100 executed (USA)	96,0000000%	(3,25)	1,75

1) Sigma Level is based on z-values of the standard normal distribution +/- 1,5s as adjustment for time-dependent performance drifts, according to Mikel J. Harry (Motorola)**
2) Sigma Level is based on z-values of the standard normal distribution according to Carl Friedrich Gauss

The appropriate target for a desired performance level varies depending on the industry, the process, the requirements of management, and the specification of customers. Naturally, an automated process in production can achieve a higher yield than the sales process in a car dealership. Ninety nine percent yield can, therefore, be excellent in one process and alarming in another. Each yield should thus be evaluated within meaningful and appropriate specification limits

Table 2.2 Adjustments to the Sigma levels according to the 1.5 standard deviation sigma shift for short-term and long-term collected data

		... and they should be compared with data measured in:	
		short-term interval	long-term interval
Data were measured in:	short-term interval	no adjustment necessary	adjustment: short-term-data - 1.5s
	long-term interval	adjustment: long-term-data + 1.5s	no adjustment necessary

values (xbar) of the samples drift about 1.5 standard deviations around the *true* mean value (μ).

This is intuitively understandable when you think of a short-term temperature measurement over the course of a week and compare them it with a long-term measurement over a year. The longer the time interval in which data is collected, the greater the probability that also larger deviations and even extreme values will occur. This increases the variability of the data and reduces the Sigma Level.

Mikel J. Harry, the architect of Six Sigma at Motorola, introduced appropriate adjustments to the Sigma levels, distinguished between measurements over a short- vs. long-time interval. The assignment to long term vs. short term depends on the actual drift of the measured values over time (Table 2.2).

The 3.4 ppm from the standard example of Motorola correspond to a z-value of 4.5 in the standard normal distribution according to Carl Friedrich Gauss. A z-value of 6 in the standard normal corresponds to 0.002 ppm or two errors per 1 billion opportunities, as shown in Table 2.1. Harry (1988) suggests maintaining this level of 4.5 sigma if data are based on long-term measurements. However, if this data is to be compared with data from a short-term interval, then 1.5 s is added. Thus, 3.4 ppm correspond to 6 sigma for short-term and 4.5 sigma for long-term measurements. The consideration of the shift allows on the one hand to estimate the change in performance of own processes over time and on the other hand to make fair comparisons with other processes and their measurements.

I did so for the example data. The example with the *Oktoberfest* data is based on a short-time measurement of 15 days. The short-term yield corresponds to a short-term sigma of 4.05 and a long-term sigma of 2.55 (-1.5 s). We could now check whether a larger proportion of drivers really lose their driving license in the course of a year, according to the shift from 4.05 to 2.55 sigma. We could also check the other examples whose measurements are based on a time interval of at least one year and to which I have added 1.5 s to compare them with corresponding imaginary short-term measurements.

But we would not need this for your project. In a typical Six Sigma project we compare short-term measurements from the measure phase before with short-term measurements from the control phase after improvements. And for both samples we will calculate the Sigma levels based on z-values, according to the standard normal distribution.

The sigmaGuide tool contains a calculator that allows you to make an initial assessment of the performance of your processes, for yield, DPMO and Sigma Level, with and without adjustments.

What is the appropriate target Sigma Level for your process? The typical level at the beginning of my projects is between −1 and +3 sigma. Nothing to be ashamed of. The right target level for your project, however, depends on your process and your industry. Of course, an automated process in semiconductor production can reach a higher yield than the sales process in a car dealership. On the other hand, process capability depends on the specifications of your customer and the requirements of your company. 99% yield can, therefore, be excellent in one process and alarming in another. In any case, however, the following applies: 6 sigma is a vision for many processes and sets the direction in which we should go. And every successful project brings us one step closer.

Questions (SC= single choice; MC= multiple choice)

Video 9/ Ch. 2.5/ #1 (SC)	From which indicator can the standard deviation already be estimated, based on the curve of a normal distribution?	
1	Skewness	
2	Kurtosis	
3	Inflection points	
4	Quartiles	
4	points	

Video 9/ Ch. 2.5/ #2 (MC)	How can the process capability index be improved?	
1	Centring of the process	
2	Reducing the scatter of the process	
3	Changing the specification limits	
4	Eliminating outliers	
4	points	

Video 9/ Ch. 2.5/ #3 (SC)	Which z-value on the x-axis of the normal distribution corresponds to the area proportion of 3.4 ppm?	
1	1	
2	1.5	
3	4.5	
4	6	
4	points	

2.6 Linking Practice, Model, and Statistics[8]

A Six Sigma project evolves at the intersection of reality, its modeling, and statistics. The reality is the source for our project topics. Here we observe the status and change of objects. This, e.g., allows us to detect differences between cookies and that

[8] Video 10 (https://www.youtube.com/channel/UCShsy50l5dFsoSXU__hBKSA)

baking time is related to different colors of cookies. Differences and relationships are the elements for our assumptions of causal relationships.

In the process model, the cookies are mapped as outputs and the timer as an influential resource in the process. To solve the problem of burnt cookies, hypotheses are formulated on the differences between outputs, and on the relationship between influences and outputs, based on our observations and assumptions. The colors of different cookies and the different settings of the timers are measured on mathematical scales. The resulting data are statistically tested according to the hypotheses (Fig. 2.13).

The chronology of the steps we take in our project is defined in the DMAIC. This is a generic scientific approach. And this approach is aimed at solving the central function of Six Sigma: Y= f(x).

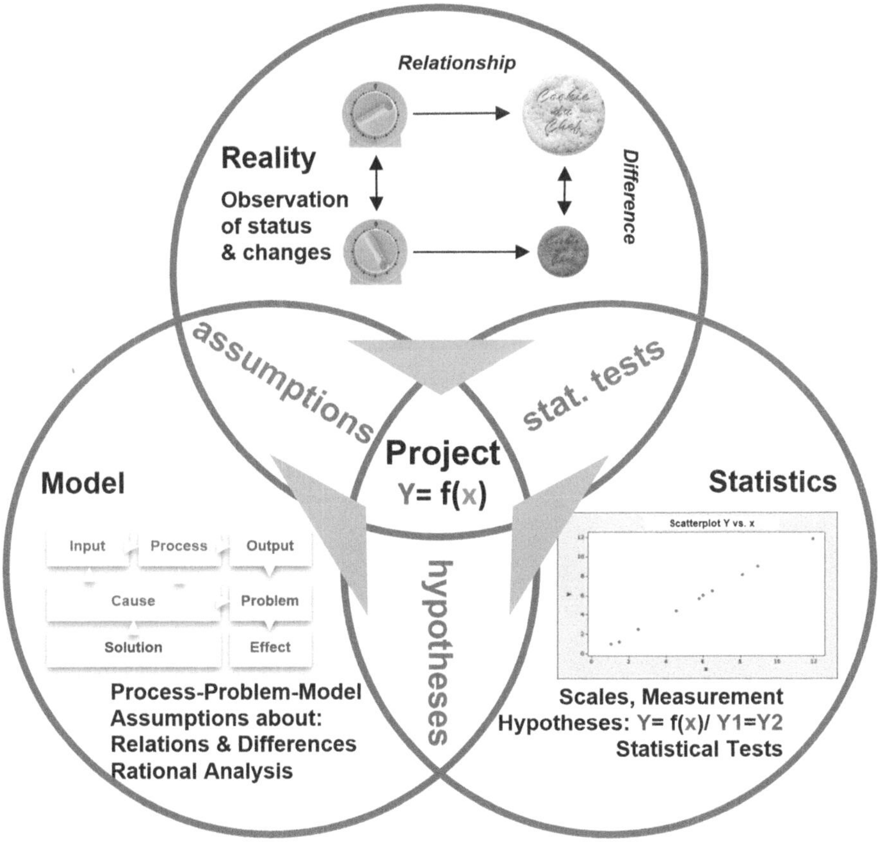

Fig. 2.13 A Six Sigma project evolves at the intersection of reality, its modeling and statistics. It is aimed to solve the equation: *Y*= *f*(*x*)—the problem is a function of its negative influences

Questions (SC= single choice; MC= multiple choice)

Video 10/ Ch. 2.6/ #1 (SC)	Typically, a Six Sigma project is used to investigate:		
		... influences (x) and problems (Y)	... problems (Y)
1	Relationships between ...		
2	Differences between ...		
2	points		

References

Harry, M. J. (1988). *The nature of six sigma quality*. Schaumburg, IL: Motorola University Press.

Leyendecker, B., Schindewolf, S., Hutwelker, R., & Weigel, H. (2011, July). Erfolgsfaktoren für die Etablierung von Six Sigma. *QZ—Qualität und Zuverlässigkeit*.

Shewhart, W. A. (1939). *Statistical method from the viewpoint of quality control*. Washington, D.C: Reprint of the Graduate School, Department of Agriculture.

sigmaGuide 3

3.1 Download and Installation

sigmaGuide helps you to implement your project. It contains all relevant logical–rational tools for the DEFINE, MEASURE, IMPROVE, and CONTROL phases. It does not contain statistical tests for the ANALYSIS phase. Therefore, you also need a statistics program for your project, e.g., Minitab.

sigmaGuide is a licensed Excel file. It is available for you as a free trial version for 100 days. Our Green Belt certification students will receive a 1-year license. Unfortunately, it actually only works under Windows operating systems (Win 7 and later). So, you need a Windows operating system to work with it. You can download sigmaGuide from my homepage (Fig. 3.1): http://www.sigmalogic.de

After your download please move the sigmaGuide.zip file to an empty folder on your local hard drive. sigmaGuide does not work on a network drive, not in the cloud and not within the ZIP archive.

Unzip the four archived files in the folder. You will find:

- **sigmaGuide.exe**: This is the executable program you must start first.
- **sigmaGuide.xlsm.aipexcel**: This is the empty version of sigmaGuide for a Business-Project.
- **sigmaGuide_Standard-Project_Environmental-Littering.xlsm.aipexcel**: This is the predefined version for the Standard-Project.
- **sigmaGuide.lic**: The license file necessary to start sigmaGuide.exe.

Video 11 (https://www.youtube.com/channel/UCShsy50l5dFsoSXU__hBKSA)

© Springer Nature Switzerland AG 2019

R. Hutwelker, *Six Sigma Green Belt Certification Project*, Management for Professionals, https://doi.org/10.1007/978-3-030-31915-1_3

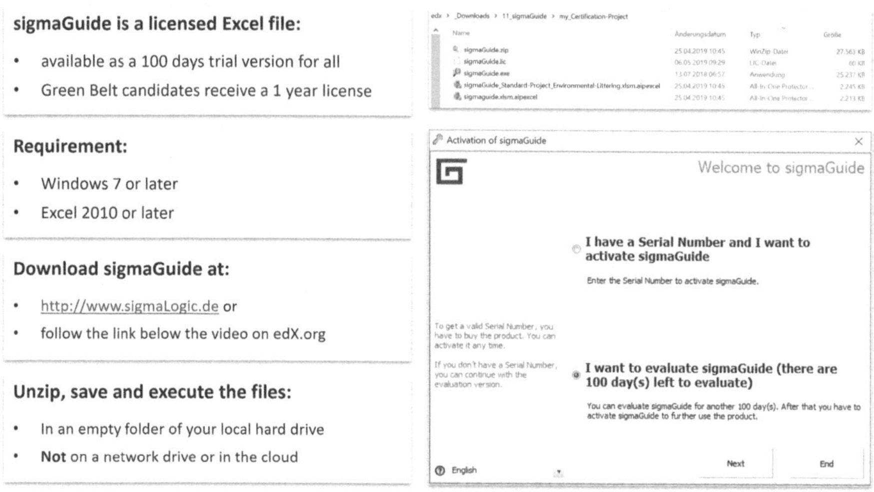

Fig. 3.1 Preparation of sigmaGuide: Unzip, install, and execute sigmaGuide in an empty folder of your local hard drive. sigmaGuide does not work inside the ZIP file, not on a network drive or in the cloud

When you start sigmaGuide.exe, a file-open box appears. Please select: **sigmaguide.xlsm.aipexcel** to open the empty version for your individual Business-Project or open **sigmaGuide_Standard-Project_Environmental-Littering.xlsm.aipexcel** to open the Standard-Project.

In the following dialog box select: *I want to evaluate sigmaGuide* and then click: next. After that Excel is started with sigmaGuide and you are ready to go.

When you work with sigmaGuide it may be necessary to relocate cell contents. However, please never use: *Cut & Paste*, as this can cause references to be lost between the cells. To shift cells always use: *Copy & Paste* and then: *Delete* unnecessary cell contents.

Before you exit sigmaGuide please always:

1. Save and then
2. Save as, under a new file name, e.g., with an increasing version number.

This allows you to track all your previous work and assures that nothing will be lost.

If you have problems opening sigmaGuide, then please first check whether you really followed all described steps. If you still have problems, please contact me (reiner.hutwelker@sigmalogic.de).

3.2 Tools and Features

sigmaGuide is based on the process-problem-solving model (Sect. 2.4). It is divided into three parts.

In **Part 1** topic for a Six Sigma project is identified. This part is aimed at all employees of the company who want to be involved in the improvement of the company's processes without prior knowledge of Six Sigma. Here, an observed weakness of the process can first be briefly described and then further detailed. This information is summarized into a final statement about the *Project-Topic*. Furthermore, the suitability of the topic for a Six Sigma project is evaluated in an index. Another indicator shows the relevance of the topic for the company.

In **part 2** the Project-Topic is further specified by describing the respective product or service, its problem is specified and its underlying process. Furthermore, any solution ideas that may already be available can be entered. The *quality*, *availability*, and *consumption* of the product/service are evaluated as well as their *effects* on the customer and on the business. This part is aimed at employees with and without previous knowledge in Six Sigma. The result—the *Project-Definition*—can serve as a basis for the management to select appropriate Six Sigma projects and to assign a Sponsor for each project.

Part 3 contains the core of the DMAIC. Here the Six Sigma project is implemented, and this part is aimed at Green and Black Belts, or Belt candidates (Figs. 3.2 and 3.3).

In all parts you will be guided by questions and receive a summary of your answers. The summary results from a controlled dialog. It can help you to evaluate the appropriateness of your answers in the respective tool. The summary can also serve as information for your coach and Sponsor about the progress of your project (Fig. 3.4).

Phase	Tool	Purpose
	Summary & Focus	Summary of the purpose and focus of the tools in sigmaGuide
Define	**Part 1**	**Identify a topic for a Six Sigma project**
	Project-Topic	Identify Problems of the daily work as a potential for Improvement
	Part 2	**Define a Six Sigma project**
	Process & Output	Describe the Process and its Output
	Problem	Describe the Problem
	Effect	Describe the Effect of the Problem
	Solution	Outline Solution-Ideas - if already available
	Project-Definition	Summary: Process, Output, Problem and Effect
Define	**Part 3**	**Implement a Six Sigma project**
	SIPOC	Structure the Process in its important steps, with related Supplier, Inputs and Outputs and Customer
	Voice to Critical	Identify Voice of Business/ Customer (VoC/ VoB), Critical Business/ Customer Requirements (CCR/ CBR), Problems and CtQ´s
	Voice to Critical (Summary)	Summary: VoC, VoB, CCR, CBR, Problems and CtQ´s
	Chart: Y CtQs Bar-Chart	Problems and their Severity
	Chart: Y CTQ Kano	Fulfillment of requirements resulting satisfaction according to Kano-characteristics
	Project-Charter	Complete and sign the Project-Charter
	Stakeholder-Communication	Identify the Stakeholder and develop a Communication-Plan
	Chart: Stakeholder	Power and interest of and influence on stakeholders
Measure	Input-Analysis	Describe Inputs (xI) of the Process, Requirements on the Inputs and Deficiencies
	Chart: Influence of xI	Display: Strength of negative Influences of the Inputs (xI) on the Outputs (Y)
	Process-Mapping and -Analysis	Describe Activities of the Process (xP), related Inputs (xI) and Outputs (Y) and negative Influences
	Chart: Influence of xP	Display: Strength of negative Influences of the Activities (xP) on the Outputs (Y)
	C&E Matrix xY	Evaluate relationships between negative Influences of the Inputs (xI) and the Activities of the Process (xP) on the Outputs (Y)
	Chart: C&E Heatmap	Display: Risks of the Influences from Inputs (xI) and Activities (xP) on the Outputs (Y)
	Data-Collection-Plan	Operationalise Measurands of Inputs (xI), Activities (xP) and Outputs (Y); Recommendation for appropriate: Charts, Parameter, Process-Capability-Indices, Control-Charts, One-Sample-Tests
	Hypotheses	Overview of all automatically generated Hypotheses, prioritized by their Risk; Recommendation for appropriate statistical Tests
	Hypotheses (Templates)	Templates to summarize the C&E Heatmap and to formulate additional Hypotheses
Analyse	Process-Capability	Calculate Process-Capability (Yield, DPMO, Pp/ Ppk, Sigma-Level, …)
Improve	Solutions	Develop Solutions to eliminate, adjust or circumvent the Root-Causes
	Chart: Solution-Selection	Overview to Efforts, Benefits and Effect of Solutions
	Action-Plan	Specify Measures to implement the Solution-Ideas
	FMEA	Analyse the Risks of Measures (FMEA:= Failure Mode and Effects Analysis)
Control	Process-Capability	Calculate Process-Capability (Yield, DPMO, Pp/ Ppk, Sigma-Level, …)
	Process-Management-Plan	Define measures to sustainably maintain the process-improvements
	Summary and Benefits	Summarize the results of the phases and demonstrate the financial and other benefits of the project
General	Lessons Learned	Summarize lessons learned and identify topics for future projects
	Belt Checklist and Evaluation	Project tasks that must be completed for certification and their evaluation by the Master Black Belt
	Sponsor Checklist	Tips and Tasks for the Sponsor

Fig. 3.2 Overview of sigmaGuide with 30 linked tools and charts and additional checklists for the Green Belt candidate, the Master Black Belt, and the Sponsor from identifying a suitable Project-Topic to evaluating the benefits of the implemented measures

Many tools guide you by a controlled dialogue:

1. sigmaGuide asks you a question
2. You enter the required information
3. sigmaGuide presents you a summary of your information in a statement
4. You check if the statement summarizes your information in a meaningful way
5. If yes - great
6. If not, please adjust your answers so that the statement makes sense.

Fig. 3.3 Controlled dialog in sigmaGuide which leads you to the right answers in several tools

Summary & Focus	S (Solutions for Causes)	X (Influences/ Causes)	Y (Problems)	Z (Effect of Problems)	Organisation, Communication, Cooperation	Description
Project-Topic			Y	Z		Identify a Project-Topic, check it for its Six Sigma suitability, show it to a potential Sponsor and ask for support.
Project-Definition	S	X	Y	Z		Identify the underlying Process, its Output, its Problems, their Effects and any existing Solutions. The information is summarized in a Project-Definition. It should be evaluated by the Sponsor/ Master-Black-Belt.
SIPOC		X	Y			The field of the Project is structured into the important Process-Steps, from the first Input to the last Output. All subsequent tools are based on these Inputs and Outputs. Keep the SIPOC up-to-date during the project.
Voice-to-Critical			Y			- SIPOC outputs are linked to Voice-of-Customer (VoC) and Voice-of-Business (VoB), - the Critical-Customer-Requirements (CCR) and the Critical-Business-Requirements (CBR) are inquired, - assigned to a Requirement category (Quality, Availability, Consumption), - the degree of their fulfillment is assessed and weighed by the criteria of the KANO-Model and - deviations from the Requirements are described as Problems. This results in a ranking order of the Outputs, their Requirements and Problems. The ranking starts with the critical Outcome variables Y, i.e. the Critical-to-Quality (CtQ). The Chart Y CTQ shows the severity of the problems in a bar chart. The Chart Y CTQ Kano shows the severity of the problems related to the Kano characteristics.
Project-Charter	S	X	Y	Z	OCC	- summarize the important information about the Process, its Problems and Business Relevance and - supplemented by information about the scope, targets, dates and Roles. Thus the mandate for the Six Sigma Project is defined.

Fig. 3.4 Summary and investigation focus for each tool in sigmaGuide

All tools of sigmaGuide are briefly described in the summary. The focus of each tool is indicated by the corresponding variables and their labels. Each of the following spreadsheets contains one of these tools and a description of its application.

In the next videos, we will apply them together. Let us go.

DEFINE: Parts 1 and 2

4

4.1 Overview: Steps, Tools, Their Purpose, and Results[1]

Of course, we start with the DEFINE phase. The purpose of the DEFINE phase is to define a critical output, identify the important unfulfilled requirements, describe its problems (Y), evaluate their monetary business relevance and define the targets, the scope and the team for the project.

When I started my Six Sigma training in the 1990s, nobody really cared about the suitability of project topics. We received a topic from the management and often enough had to realize that it was not suitable for Six Sigma. Of course, you can almost always use one or the other DMAIC tool, but a Six Sigma project requires at least the use of the basic Six Sigma tools in their logical order, as we will do here. I wasted a lot of time on unsuitable projects and I have the impression that I am an expert on Six Sigma dead ends and project abysses. Therefore, I developed the tools Project-Topic and Project-Definition. They allow to involve every employee of a company in the identification of suitable project topics (Fig. 4.1).

Thus, at first, we will identify and define a new Six Sigma project with the tools: Project-Topic and Project-Definition. Then we will delimitate and structure the application area of the project in the SIPOC. We will listen to the customer and the responsible manager (Voice of Customer, VoC/Voice of Business, VoB), filter their complaints and requirements (Critical Customer Requirements, CCR/Critical Business Requirements, CBR), prioritize the problems (KANO-Model) and identify problems that are Critical to Quality (CtQ). At least we will define the scope of the project, define its targets, build the team, and agree these conditions in the Project-Charter.

For the Project-Topic we have to identify a weakness in a process and/or a deviation in a product or service, evaluate its relevance for the company and check the suitability for Six Sigma (Project-Topic).

[1] Video 12 (https://www.youtube.com/channel/UCShsy50l5dFsoSXU__hBKSA)

© Springer Nature Switzerland AG 2019

R. Hutwelker, *Six Sigma Green Belt Certification Project*, Management for Professionals, https://doi.org/10.1007/978-3-030-31915-1_4

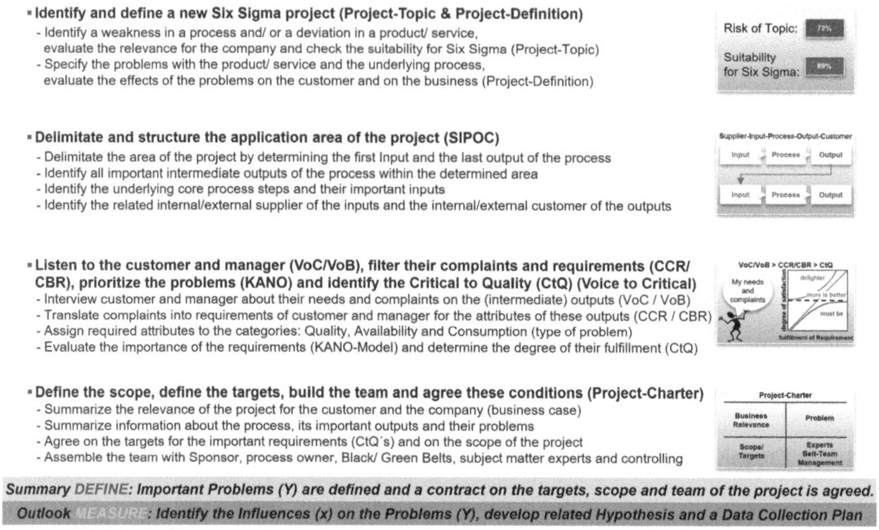

Fig. 4.1 Overview DEFINE: From the identification of problems (Y) to the Project-Charter

Then we will specify the problems with the product/service, identify the underlying process, and evaluate the effects of the problems on the customer and on the business initially from our perspective (Project-Definition).

In the SIPOC we will delimitate the area of the project by determining the first input and the last output of the process. Then we will identify all important intermediate outputs of the process within the determined area as well the underlying core process steps and their important inputs. At least we will identify the related internal/external supplier of the inputs and the internal/external customer of the outputs.

I combined the tools: Voice of Customer, Voice of Business, the identification of critical requirements, the KANO-Model, and the CtQs in one tool and called it: Voice-to-Criticals. We start by interviewing customer and manager about their needs and complaints on the (intermediate) product/service (VoC/VoB). Then we translate the complaints into critical requirements for the attributes of the product/service (CCR: Critical Customer Requirements/CBR: Critical Business Requirements). Next, we assign the critical attributes to the categories: *Quality*, *availability*, or *consumption* and thus determine the type of the problem. At least we evaluate the importance of the requirements according to the KANO-Model and determine the degree of their fulfillment, resulting in the CtQs—Critical to Quality.

In the Project-Charter we will summarize the relevance of the project for the customer and the company in a business case. We also summarize information about the process, its important outputs and their problems, and agree on the targets for the important requirements (CtQs) and on the scope of the project. Now the necessary team can be assembled, with a Sponsor and a responsible manager of the process, which is often the same person, a Green or Black Belt (candidate), the necessary experts and a person from the (financial) controlling department, who ultimately reviews the benefits of the project.

All important problems (Y) are now defined and a contract on the project is agreed. After DEFINE we can go to the MEASURE phase to identify the influences (x) on the problems (Y), develop related Hypothesis and a Data-Collection-Plan.

4.2 Project-Topic[2]

4.2.1 Identification of a Project-Topic

Mike Rother and John Shook published their book: Learning to see, about Value-Stream Mapping to Create Value and Eliminate Muda in 2003. *Although I do not want to talk about value-stream mapping here, I want to emphasize the importance of the title "learning to see" for any kind of process analysis and process improvement.*

Things we often see are taken for granted, and changes are only perceived when we can no longer assimilate them to the normal state. We are sometimes so generous with deviations that we miss the point, at which we can still take corrective action. When we recognize deviations, we sometimes even blame them for causes that did not even exist. The causal argument: The cookies of our bakery tasted bad because our chef did not use a recipe, is one of them. Please think about this argument.

We will later return to this topic in the Root-Cause-Analysis. But please note already here that during the identification of a Project-Topic, as with all other tools, it is important to open all our senses and reproduce our impressions as precisely as possible.

Open sigmaGuide and go to the tool: Project-Topic. We switch into the role of an employee of our bakery and answer its seven questions to give a short description of the observed weakness (Fig. 4.2).

1. The first question asks to describe the observed weakness in a short statement. It is not necessary here to describe the causes of this problem or any solution ideas. Just describe the weakness in everyday language. I observed that:

   "The bakery cookie du Chef has several problems. Most important: sometimes the cookies taste bad. Sometimes the cookies are delivered too early. Additionally, there is some waste of ingredients and energy."

   If you like, then take some time now to identify a weakness in your work environment and briefly describe it in your own words.

[2]Video 13 (https://www.youtube.com/channel/UCShsy50l5dFsoSXU__hBKSA)

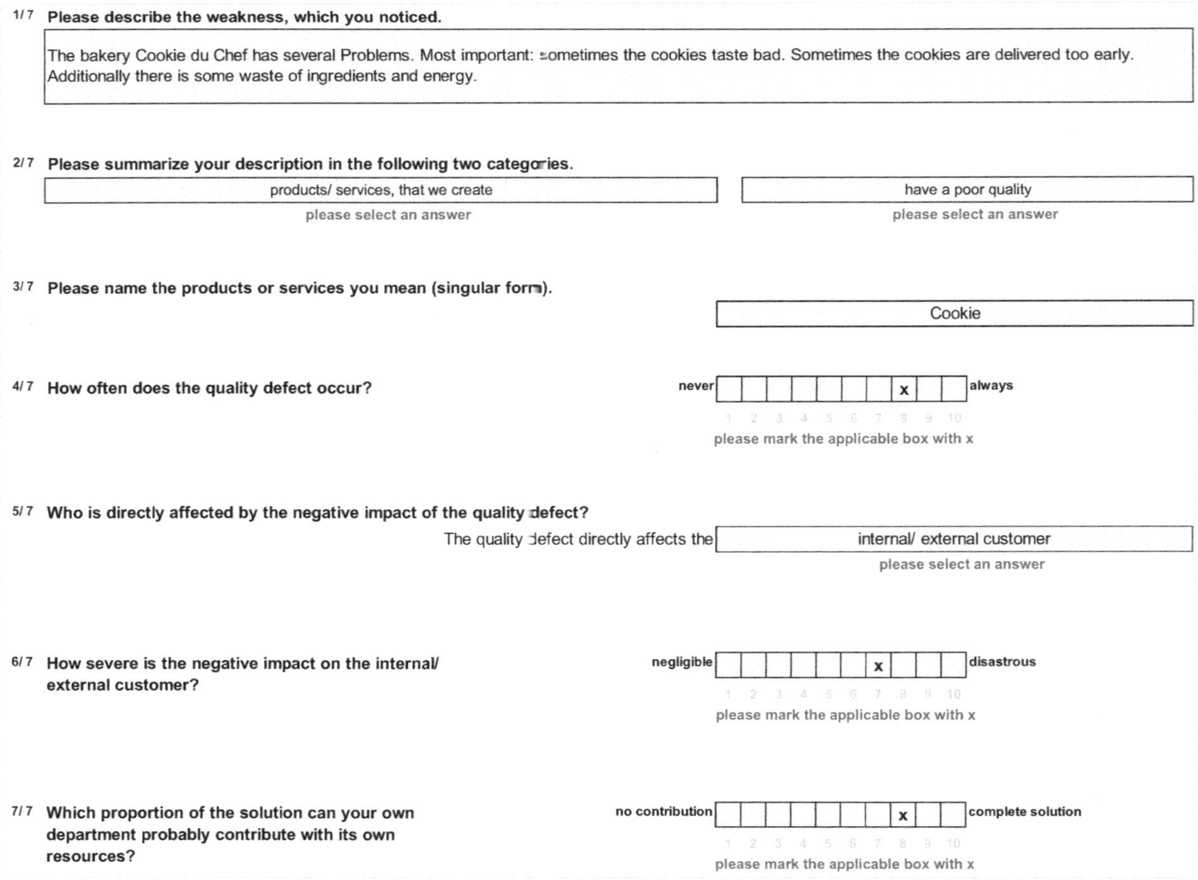
1/7 **Please describe the weakness, which you noticed.**

The bakery Cookie du Chef has several Problems. Most important: sometimes the cookies taste bad. Sometimes the cookies are delivered too early. Additionally there is some waste of ingredients and energy.

2/7 **Please summarize your description in the following two categories.**

products/ services, that we create — please select an answer

have a poor quality — please select an answer

3/7 **Please name the products or services you mean (singular form).**

Cookie

4/7 **How often does the quality defect occur?**

never [ ][ ][ ][ ][ ][ ][ ][x][ ][ ] always

1 2 3 4 5 6 7 8 9 10

please mark the applicable box with x

5/7 **Who is directly affected by the negative impact of the quality defect?**

The quality defect directly affects the internal/ external customer

please select an answer

6/7 **How severe is the negative impact on the internal/ external customer?**

negligible [ ][ ][ ][ ][ ][ ][x][ ][ ][ ] disastrous

1 2 3 4 5 6 7 8 9 10

please mark the applicable box with x

7/7 **Which proportion of the solution can your own department probably contribute with its own resources?**

no contribution [ ][ ][ ][ ][ ][ ][ ][x][ ][ ] complete solution

1 2 3 4 5 6 7 8 9 10

please mark the applicable box with x

Fig. 4.2 The identification of a Project-Topic is the first step to a Six Sigma project. The video: Cookie du Chef shows (Sect. 2.3) that there are improvement potentials in this bakery. Every employee is able to describe observed weaknesses in the work environment

The next questions and their predefined answer choices are intended to describe the weakness in more detail, to determine whether the topic is suitable for a Six Sigma project.

2. The second question asks to summarize the description in two categories. You can select between:
   (a) Tasks or orders that we receive
   (b) Information or material that we receive
   (c) Rules, instructions, or routines that direct how we work
   (d) Devices, equipment, or software that we use
   (e) Activities we perform and
   (f) Products or services that we create.

   These categories are based on the components: input, method, resources, activities, and output of the integrated process-problem-solving model (Sects. 2.4.2 and 2.4.3) to determine the critical object of the observed weakness. Our topic concerns the bad taste of the cookies and thus I select: products or services that we create.

   The second part of this question asks how the object deviates from the expectation. Here you can select between:
   (a) Lead to waste/are no longer needed
   (b) Have a poor quality and
   (c) Take a long time/are available too late.

   The bad taste of the cookies is of course a sign of: Poor quality.
3. The third question asks to specify the product or service and I simply enter: Cookies.
4. Because the cookies nearly always taste bad, I put my mark at level 8 for the occurrence of this quality defect.
5. Question five asks for the negatively affected subject of the: Poor quality and offers the options:
   (a) Internal/external customer
   (b) Employees
   (c) Supplier/service provider/partner and
   (d) Own department/our company.

   Although I also like cookies, it is primarily the: External customer who is affected from our weakness.
6. In question six I rate the degree of direct negative impact on the customer quite high with a 7.
7. Since the bakery is under the responsibility of our Chef and her manager, the solution is in our hands and I select 8 in question seven to indicate our possible contribution to future solutions.

Now the summary of my answers is given in three statements. At first there is a copy of the description of my observation (Fig. 4.3).

The second statement is:

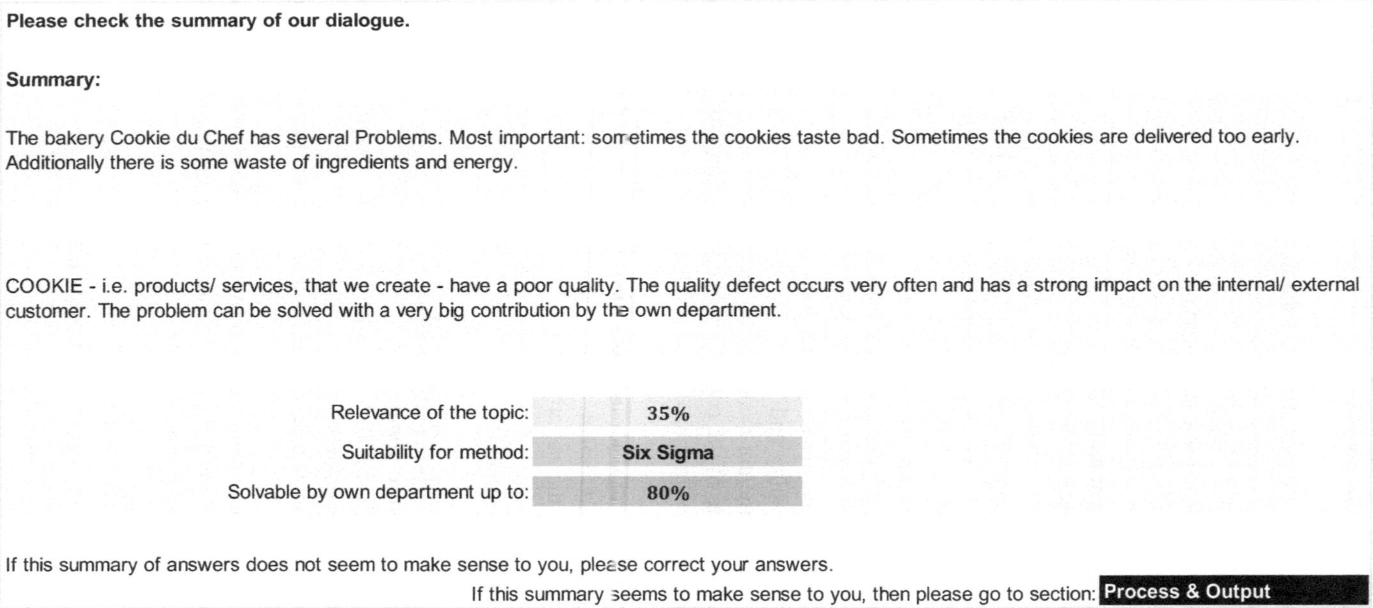

Fig. 4.3 The Project-Topic summarizes the given information in statements and evaluations. Based on this summary, the management can decide whether further investigations are necessary

Cookies—i.e., products/services that we create—have a poor quality. The quality defect occurs very often and has a strong impact on the internal/external customer. The problem can be solved with a very big contribution by their own department.

This statement summarizes the questions and my answers on the topic. The sentence can be read fluently, and I find that the statement describes the situation in our bakery.

In the following tools there are more such sentence templates whose gaps are filled with the given answers. If one of these summaries does not make sense, sounds very cumbersome, contains duplicates, or is grammatically incorrect, you may need to reformulate one or more of your given answers to get a more appropriate or precise statement.

The third section evaluates the answers and offers an interpretation of the:

- Relevance of the topic for the company and the
- Suitability of the topic for Six Sigma.

Alternative suggestions for your Project-Topic might be:

- Lean Six Sigma, if you have a problem in the availability of your product or service or
- Basic Improvement, if you observed waste of input or resources.

The final percentage value repeats to what extent the problem can be solved with the resources of your own department.

4.2.2 Criteria for Lean Six Sigma Project-Topics

On this first stage of project identification, we would overwhelm our colleagues with an in-depth evaluation of the topic for Six Sigma suitability. You will find several lists of criteria in the web.

The important criteria for me are:

1. The focus is on the deviation of an output and not an input. If the weakness is a deviation of the input, it would be a project for the (internal) supplier. Deficiencies primarily in the method or resources are probably more of a: Basic Improvement topic.
2. A quality deviation is an indicator for Six Sigma. Waste or reduced availability is more a Lean or a Basic Improvement topic.
3. The effect of the problem on the external customer is also a criterion for Six Sigma. But we would also accept negative effects on internal customers.
4. Another important factor is: Measurability of the output problem (Y) and its triggering influences (x) in the process. For a *Six Sigma* project, both the influences (x) and the problem (Y) should be measured and data collected. This is often the case with quality problems.

Availability and consumption problems are topics for Lean/Kaizen, if only the output performance is measured. Basically, time losses and waste of resources can also be measured. However, in my experience, these losses and wastes often occur in all process steps—in contrast to influences on quality, which occur more concentrated. In many cases it would be very costly to measure these many influences and often it is not even necessary. We know that we should reduce waste and loss. And at least in Germany it is rightly forbidden to measure employee performance.

If it is not or cannot be measured at all, then it is a topic for: *Basic Improvement* or it is simply *Actionism*.

5. If influences (x) and problems (Y) can be measured, it is also necessary that a sample of at least 30 x–Y value pairs can be drawn within 1 month to be able to obtain relevant and significant results from statistical tests. This means that the process must have a certain repetition rate and at least one output should be generated daily. I remember the production of an online catalog that was updated quarterly. Of course, you can also measure different output attributes and their influences in this process. However, it would have taken several years to obtain a sufficiently large sample for the statistical tests. And the use of statistical tests is essential for our Six Sigma certification project.
6. The certification project should also bring financial benefits to the company and be
7. Implemented within 6 months.
8. During this time, it should be possible to interview customer and involve experts in the workshops.

Despite the weaknesses of the evaluation and the resulting draft interpretation, the summary of the Project-Topic might be overall helpful for the responsible manager, to get a first impression about the topic and to decide on further investigations. We will do this in the following Project-Definition and apply some more of the presented criteria to the Project-Topic (Fig. 4.4).

Questions (SC= single choice; MC= multiple choice)

Video 13/ Ch. 4.2/ #1 (SC)	Please assign these influences, types of problems and the degree of measurability to the different project types.			
		Basic Improvement	Lean Sigma	Six Sigma
1	Method wrong/ incomplete/ not available			
2	Resource defect/ not available			
3	Activity laborious/ inefficient/ repeated			
4	Output defect/ not available			
5	Quality problem			
6	Availability problem			
7	Consumption problem (waste)			
8	x and Y measurable			
9	only Y measurable			
10	nothing measuable			
10	points			

Tasks

Business-Projects: Please identify a Project-Topic and enter the required data in the tool.

Focus: Problem of the Output
Type of problem: Deviation in Quality, Availability and Consumption
Effect of the problem on: (internal) Customer and Business

Measurability:

x (influences of Input/ Method/ Resources) \ Y (required attributes of the Output)	measurable	not measurable
measurable	Six Sigma	really?
not measurable	Lean/ Kaizen	Basic Improvement or Actionism

Statistical tests: Sample size > 30 xY value pairs collectable within a month
Project benefit: 20 - 60k€ for a Green Belt and > 100k€ for a Black Belt project
Implementation time: Not longer than half a year (typically)
Persons to be involved:
- Customer for interviews
- Experts for workshops **(Process-Mapping; Root-Cause-Analysis & Solutions)**
- Employees to implement the developed measures

Fig. 4.4 Criteria for Lean Six Sigma Project. For the Lean Six Sigma certification you must solve a quality, availability, and consumption problem

Standard-Project: For your orientation the tool Project-Topic is already pre-filled. The entries result from the perspective of a garbage disposal employee. Imagine that you are employed by the garbage disposal service and responsible for the cleanliness of streets and squares. Probably there are more attractive employers for you, but if you cannot do a business project, then this project offers the opportunity to get the Green Belt certificate and do something positive for the environment.

4.3 Project-Definition[3]

The challenge in the tool: Project-Topic was to identify a weakness of a process or product. Further information is now collected in the Project-Definition to specify the topic.

The management of our bakery decided to further explore the Project-Topic. Typically, they would ask a Green or Black Belt to specify the process and output of this topic, specify the problems and their effects on customer satisfaction and costs, collect already available solution ideas and summarize the information in a Project-Definition. Thus, I received the Project-Topic and checked the given information. Then I conducted an interview with the author of the Project-Topic and asked the following questions.

[3] Video 14 (https://www.youtube.com/channel/UCShsy50l5dFsoSXU__hBKSA)

4.3.1 Process and Output

First, I need to know the name of the process in which the cookies are generated. I often experienced that companies name their processes with just one noun. I prefer to label processes with a verb and a noun, to specify the main activity of the process, followed by the object of the activity. Thus, I enter: "bake cookies" to denominate the main process of our bakery (Fig. 4.5).

The second question asks: "Which type of output is: Cookie?" Sometimes it might not immediately be clear whether an output is a material product or an intangible service. Cookies are of course tangible products. Although there is no difference in the procedure of Six Sigma between products and services, it might be helpful for the management to distinguish the type of output for their selection of the most promising project.

Question 3 asks for the status of the output and its customer. The dough, e.g., is an intermediate output for internal customer and a cookie is a final product for external customer.

Question 4 evaluates the type of process. The possible answers distinguish between:

- Development processes
- Creation processes
- Logistic processes
- Operation of systems (e.g., of a data center or a chemical factory)

Baking cookies is a repeated creation of similar products. If we, e.g., would invent new recipes, then I would select: Development process of new services.

The related question 5 asks how often the product/service is generated within a year. Here you can select between different frequencies or select the continuous generation. Our Chef bakes once a week and thus I select 13–52 times. If it was about operating a data center or the continuous output in a chemical factory, then I would choose: Continuous.

Question 6 finally asks for important inputs of the process to get a better understanding of the related upstream processes.

A summary again closes this section (Fig. 4.6):

Does this summary make sense and map the reality? I think yes—and continue my interview with the description of the problems.

4.3.2 Problems

In this section the three most important problems of the product/service can be specified (Fig. 4.7).

The first question asks in which category the cookies differ from the requirements. According to the Project-Topic, the quality deviation is most important and so I choose the answer:

1/6 In which Process is the Output COOKIE generated?

Please describe the Activity of the Process with a Verb and a Noun (e.g. weigh Ingredients, knead Dough or bake Cookies).

bake cookies

please enter a Verb and a Noun

2/6 Which type of Output is: COOKIE?

Outputs can be distinguished as tangible Products (e.g. Ingredients, Dough, Cookie) or as intangible Services (e.g. Request, Order, Address, Delivery, Invoice and Remittance).

tangible Product

please select an answer

3/6 What is the status of the Product: COOKIE and who is the recipient?

Is the Product: COOKIE an intermediate Product for internal Customers within the organization or a final Product for external Customers?

final Output for external Customers

please select an answer

4/6 Which type of Process is: BAKE COOKIES?

Processes can be distinguished in: Development Processes, Creation Processes, Logistic Processes and Operation Processes. BAKE COOKIES is a Process for the:

repeated creation of similar Products/Services

please select an answer

5/6 How often is the Product: COOKIE within a year generated?

13 - 52 times

please select an answer

6/6 Which important Input is necessary for the Process BAKE COOKIES?

Which important Input is necessary for the Process BAKE COOKIES to generate the Product COOKIE?

Butter, Sugar, Flour, Chocolate

please enter your answer

Fig. 4.5 Project-Definition: questions and answers to the process and its output

Summary:

The Product COOKIE is a tangible final Output for external Customers and is in the Creation Process BAKE COOKIES within a year 13 - 52 times generated. Important Input of the Process to generate the Product COOKIE is: BUTTER, SUGAR, FLOUR, CHOCOLATE.

Fig. 4.6 Summary to Section 1: process and output

- *Wrong, defect, bad, or faulty.*

As you can see the answer:

- *Not or only partially suitable for the targeted purpose* also belongs to the category of: Quality.

The next three answers options:

- *Not available in the right quantity (too much/too few)*
- *Not available at the right time (too early/too late)*
- *Time consuming in preparation or waiting times in between*

describe the: Availability of an output.

The last answer option: *wasteful in consumption of input or resources* of course belongs to the category of: Consumption.

The second question now asks to specify the selected answer with my own words: "What exactly do I mean with: COOKIE is wrong, defect, bad, or faulty?" And I am asked to specify the affected attribute of the output and its deviation. The answer is very easy here: "taste bad." Do you remember that requirements always relate to attributes of products/services? If we want to be specific, we would not simply answer: "cookie bad" because this is a very general answer. The *taste* is the critical attribute of the cookies here and the *Taste is bad* describes the deviation of this attribute.

The third question asks how well the product/service fulfills this requirement. And I mark it with a 3.

The summary now offers the statement:

1. Problem: COOKIE TASTE BAD. COOKIE fulfills the requirement on Quality (is error-free) in 30%.

This summary makes sense to me and my interview partner and I switch to the next problem. Again, there are three questions and I describe the availability problem here: Cookies are *not available at the right time*, meaning here that the "Delivery is too early." And I rate the fulfillment of the requirement on the availability with a 7.

Finally, I specify the waste of input as the third problem: "COOKIE INGREDIENTS WASTED. COOKIE fulfills the requirement on efficient utilization of means (no waste of Input, Resources) in 50%."

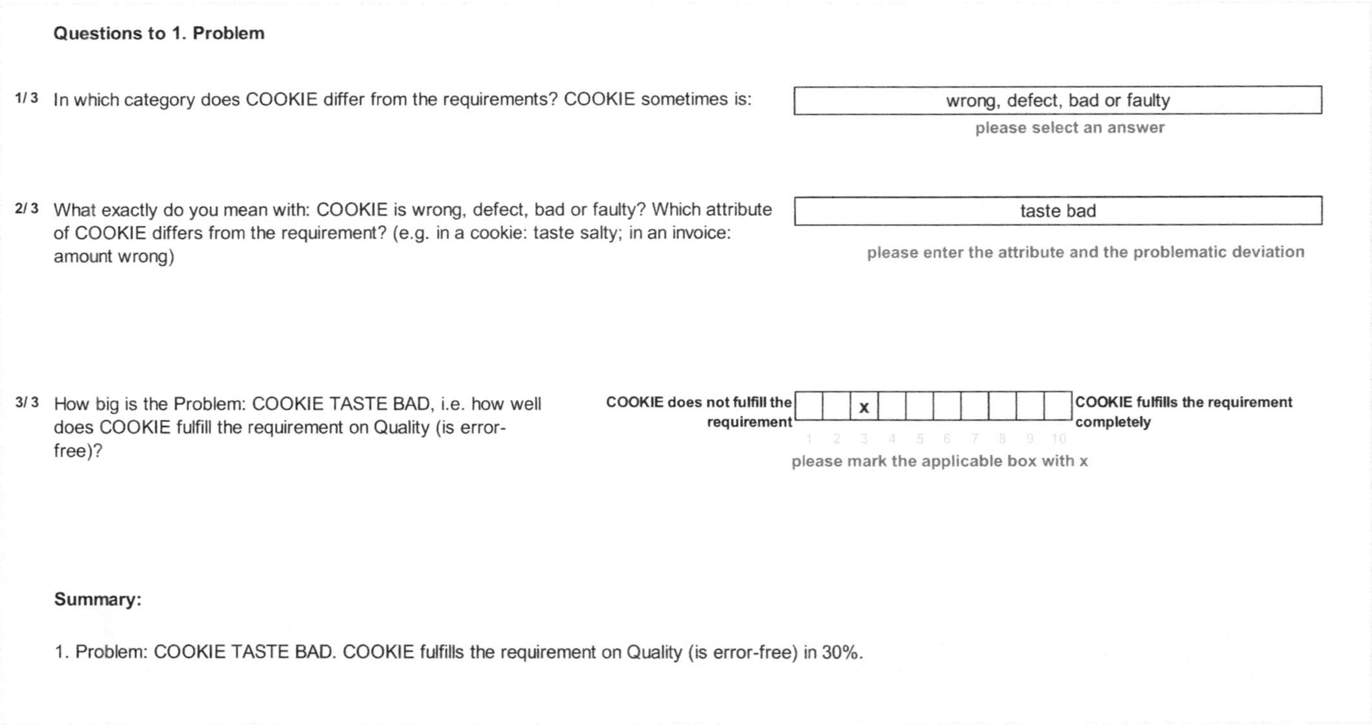

Questions to 1. Problem

1/3 In which category does COOKIE differ from the requirements? COOKIE sometimes is: wrong, defect, bad or faulty

please select an answer

2/3 What exactly do you mean with: COOKIE is wrong, defect, bad or faulty? Which attribute of COOKIE differs from the requirement? (e.g. in a cookie: taste salty; in an invoice: amount wrong) taste bad

please enter the attribute and the problematic deviation

3/3 How big is the Problem: COOKIE TASTE BAD, i.e. how well does COOKIE fulfill the requirement on Quality (is error-free)?

COOKIE does not fulfill the requirement	1	2	3	4	5	6	7	8	9	10	COOKIE fulfills the requirement completely
			x								

please mark the applicable box with x

Summary:

1. Problem: COOKIE TASTE BAD. COOKIE fulfills the requirement on Quality (is error-free) in 30%.

Fig. 4.7 Questions to Section 2: problems

4.3.3 Effects

Section 3 asks for the: Effects of the problems on the customer and on the business. Both effects are important for the evaluation of the topic and its ranking, relative to other Project-Topics.

I interviewed three customers and they rated their overall satisfaction with the quality of the cookies with a 2 and the availability of the cookies with a 4. Both answers are part of the Voice of Customer (VoC) and are summarized as (Fig. 4.8):

> The satisfaction of the external customer with the: Quality of COOKIE is: 20% and with the Availability of COOKIE is: 40%.

Additionally, I asked the process owner and her manager the questions three to seven to the effects of the problems on the company. This is part of the Voice of Business (VoB) (Fig. 4.9).

Here the summary states:

> The satisfaction of the process owners with the consumption in the creation process of the COOKIE is: 40%. The total costs of the specified three problems are estimated by 100€/year. They are primarily the result of quality costs due to scrap and additional expenditure. A solution of the problems is rated as: major URGENT (80%-Level) and major IMPORTANT (80%-Level).

If you have asked the right people and the answers make sense also in these summaries, then switch to the next section.

4.3.4 Solution Ideas

Most employees who observe and describe problems in their work environment also develop ideas about their solution. It is neither necessary to develop solutions ideas in this early stage of a project, nor is it mandatory to implement these solutions later in the IMPROVE phase (Fig. 4.10).

But there are good reasons to collect available solution ideas already here. If an employee has a solution idea that we later implement, then we should know this in advance and recognize the idea later verbally and monetary. I experienced that skeptics of a Six Sigma project pretended after the successful implementation of a solution that they already knew the solution before the project started and that my project was thus a waste. I also experienced skeptics who told me after the completion of a project, which was not as successful as expected, that they had forecasted the failure of the implemented solution. In both cases it might be helpful for you to know the existing solution ideas already at this stage to prevent fault finding.

Questions to the effects on external customer (Voice of Customer)

1/7 How satisfied are the customers overall with the Quality of COOKIE?

very unsatisfied	1	2	3	4	5	6	7	8	9	10	very satisfied
		x									

please mark the applicable box with x

2/7 How satisfied are the customers overall with the Availability of COOKIE?

very unsatisfied	1	2	3	4	5	6	7	8	9	10	very satisfied
				x							

please mark the applicable box with x

Summary:

The satisfaction of the external customers with the:

- Quality of COOKIE is: 20%.
- Availability of COOKIE is: 40%.

If this summary of answers does not seem to make sense to you, please correct your answers.

If this summary seems to make sense to you, then please scroll down to effects on the company (Voice of Business)

Fig. 4.8 Questions to Section 3: effects on the customer

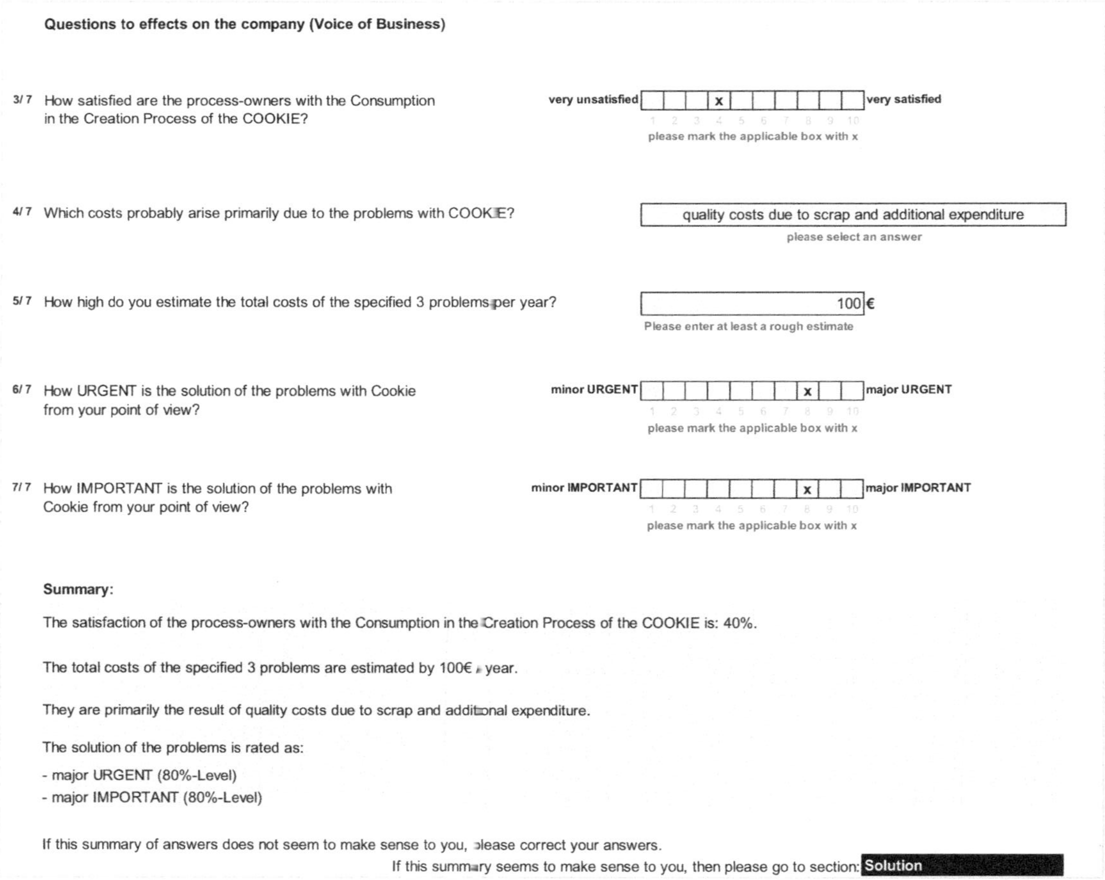

Questions to effects on the company (Voice of Business)

3/7 How satisfied are the process-owners with the Consumption in the Creation Process of the COOKIE?

very unsatisfied [ ] [ ] [ ] [x] [ ] [ ] [ ] [ ] [ ] [ ] very satisfied
1 2 3 4 5 6 7 8 9 10
please mark the applicable box with x

4/7 Which costs probably arise primarily due to the problems with COOKIE?

quality costs due to scrap and additional expenditure
please select an answer

5/7 How high do you estimate the total costs of the specified 3 problems per year?

100 €
Please enter at least a rough estimate

6/7 How URGENT is the solution of the problems with Cookie from your point of view?

minor URGENT [ ] [ ] [ ] [ ] [ ] [ ] [ ] [x] [ ] [ ] major URGENT
1 2 3 4 5 6 7 8 9 10
please mark the applicable box with x

7/7 How IMPORTANT is the solution of the problems with Cookie from your point of view?

minor IMPORTANT [ ] [ ] [ ] [ ] [ ] [ ] [ ] [x] [ ] [ ] major IMPORTANT
1 2 3 4 5 6 7 8 9 10
please mark the applicable box with x

Summary:

The satisfaction of the process-owners with the Consumption in the Creation Process of the COOKIE is: 40%.

The total costs of the specified 3 problems are estimated by 100€ / year.

They are primarily the result of quality costs due to scrap and additional expenditure.

The solution of the problems is rated as:

- major URGENT (80%-Level)
- major IMPORTANT (80%-Level)

If this summary of answers does not seem to make sense to you, please correct your answers.

If this summary seems to make sense to you, then please go to section: Solution

Fig. 4.9 Questions to Section 3: effects on the business

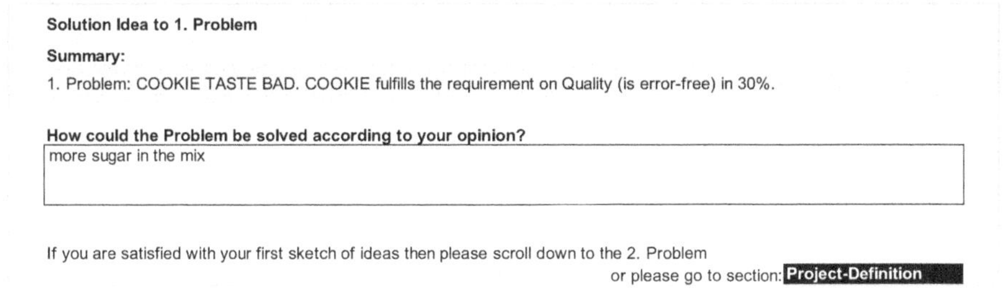

Solution Idea to 1. Problem

Summary:

1. Problem: COOKIE TASTE BAD. COOKIE fulfills the requirement on Quality (is error-free) in 30%.

How could the Problem be solved according to your opinion?

more sugar in the mix

If you are satisfied with your first sketch of ideas then please scroll down to the 2. Problem

or please go to section: Project-Definition

Fig. 4.10 Questions to Section 4: already existing solution ideas

4.3.5 Project-Definition

The Project-Definition is the summary of all previous sections with the: Project-Topic, the Process and its Output, the Problems, their Effect on the Customer and on the Business, supplemented by two charts and the first solution ideas. Nothing new is added here, except your optional comments and your personal data.

Please note: Most cells in the Project-Definition are locked, just as many other cells in sigmaGuide, which only contain references to the entries in the preceding tools. The locking is necessary because sigmaGuide gradually records information that is reused in other tools as single values or within summary templates. Changes to the locked cells would lead to inconsistent results between the tools.

If you want to change the content of a locked cell, you must do so at the source. For example, if you do not like the summary of a problem, it may be that the output is wrongly defined in the table: Project-Topic or that the relevant attribute of the output and its deviation does not match the requirements in the tool: Problem. Because a problem in sigmaGuide is composed of: Output, Attribute, and Deviation (e.g., *Cookie Taste Bad*). If you, e.g., enter a verb instead of a noun as an output, then the sentence in a summary no longer makes sense (Fig. 4.11).

I think my Project-Definition captures the situation of our bakery quite well. Now it is your turn again. Create your own Project-Definition. And then look for a Sponsor. If you get the necessary support, then document the current status of your project in your Project-Story-Book (Chap. 5). Because with the SIPOC (Sect. 6.1) we will start the core of the DEFINE phase and implement the project.

Tasks

Business-Projects: Please define your project and enter the required data in the tool.

Standard-Project: For your orientation some entries of the tool Project-Definition are already prefilled. You can customize these entries according to your needs. In any case, you must fill in the empty fields yourself. This applies, for example, to the costs incurred by the garbage collection service for the disposal of garbage on streets and squares outside the trash cans.

For example, the Federal Office for the Environment/Bundesamt für Umwelt BAFU (2011) in Switzerland calculates the annual cleanup costs of over CHF 192 million to the thoughtless littering of garbage in public areas, corresponding

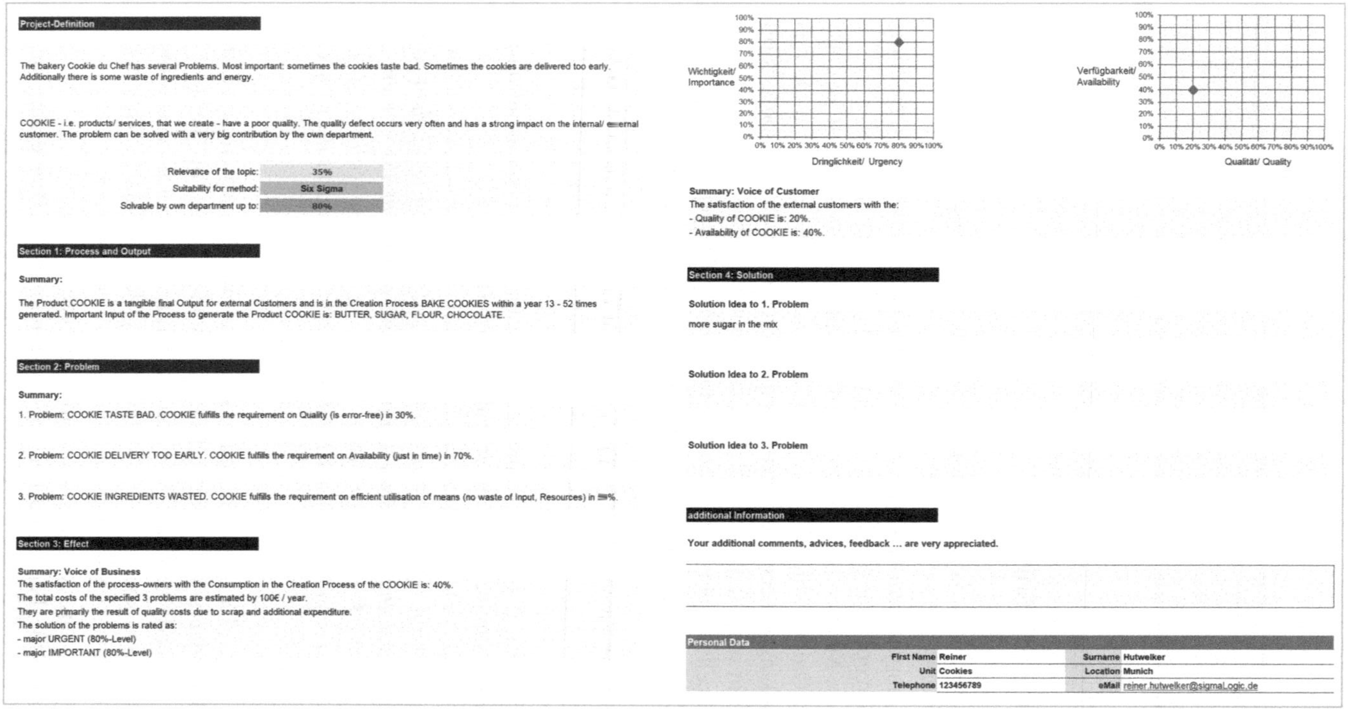

Fig. 4.11 Project-Definition: summary of all collected information about the topic, process and output, problems, effects, available solution ideas and some details about the author—ready for evaluation by the Six Sigma Board

to about $23 per inhabitant per year (https://www.igsu.ch/files/bafu_litteringkosten_2011.pdf).

Maybe there is a similar statistic for your country. The World Waste Platform offers statistics and initiatives from many countries of our planet (https://opendata.letsdoitworld.org/#/download). Or you may even get a detailed cost breakdown of your hometown from your City Hall. But since I will recommend you later in the DEFINE phase to introduce your project to responsible managers for waste disposal of your hometown, you better wait until it is time for these tools.

Reference

Rother, M., & Shook, J. (2003). *Learning to see—Value-stream mapping to create value and eliminate Muda*. Cambridge: The Lean Enterprise Institute.

Link

Costs of littering: Federal Office for the Environment/Bundesamt für Umwelt BAFU. (2011). **https://www.igsu.ch/files/bafu_litteringkosten_2011.pdf** (contains English Translation).

Project-Story-Book 5

Before we discuss further DMAIC tools, I would like to present the template for documentation of your project. This Project-Story-Book is helpful for everyone who wants to implement a project and document its steps and success. And it is mandatory for all who want to acquire the Green Belt certificate.

The template contains all necessary tools and primarily represents the results of the sigmaGuide and Minitab. It is not only about simply inserting the results, but also about highlighting important details and especially the relevant anomalies to be able to set priorities. Once you have grasped the core of a result, you can also interpret its meaning and draw conclusions for the next steps of your project.

And if you can professionally present each tool and its concrete use in your project in a slide, then your Sponsor can have a qualified say and make necessary decisions. And your Master Black Belt can see whether you can use the respective tool sensibly, that you know its purpose and conditions for use and understand the meaning and implications of its results. Thus, the Project-Story-Book also becomes the basis of our coaching.

In the content slide please add the corresponding page numbers to the tools. You are welcome to insert the name of your company and its logo in the upper right corner of the slide master as well as your name and the project title in the bottom line of the slide master.

In the next slide, please introduce yourself and your Sponsor. A photo of you, the Sponsor, and a screenshot of the company's website would also be nice here. To document that this is a real project, we also ask you to scan and insert the confirmation mail of your Sponsor with his or her signature. Finally, please introduce the critical product/service and its problem, if possible, also illustrated in one or more photos.

All project slides have the same structure (see Fig. 5.1). The text in the upper left corner contains the name of the DMAIC phase and the name of the respective tool.

Video 15 (https://www.youtube.com/channel/UCShsy50l5dFsoSXU__hBKSA)

© Springer Nature Switzerland AG 2019

R. Hutwelker, *Six Sigma Green Belt Certification Project*, Management for Professionals, https://doi.org/10.1007/978-3-030-31915-1_5

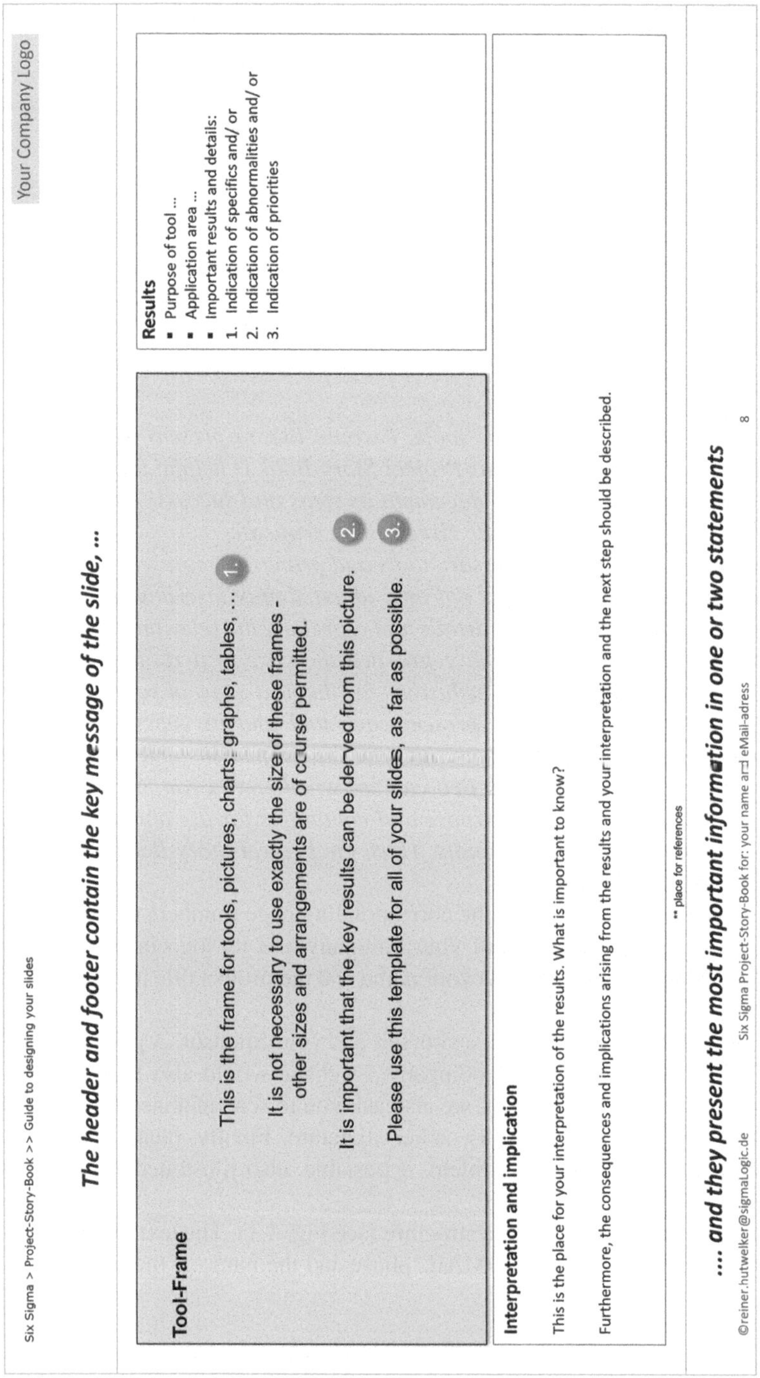

Fig. 5.1 Project-Story-Book: template for structuring and presenting the results from sigmaGuide and the statistical tests (download at: www.sigmalogic.de)

Header and footer summarize the core message of the slide in one or two statements. Every participant of the project team and every manager should be able to understand this important information immediately and without help. The Tool-Frame is the placeholder for tool results, images, charts, graphs, and tables. The Results-Box on the right should briefly summarize the purpose of the tool, the area in which the tool was used, as well as its important results and details.

To improve orientation, you can insert numbered bullets at the significant points of the Tool-Frame and explain these details in the Results-Box. The Interpretation-and-Implication-Box contains your interpretation of the results. Your implications should show the consequences of your conclusions on how to proceed. It is not always easy to get to the heart of the matter in such a way that a layman could understand the slide. But this is necessary. It requires close observation and reduction to the core of the observation.

The Project-Story-Book formally mirrors my Six Sigma slides and hopefully offers you many useful clues to insert the specific contents of your project.

Special cases are the slides for the two workshops. In the MEASURE phase you will map the process and identify the triggering influences on the problem. This is best achieved with a workshop with all involved employees of the process. However, it makes no sense to enter the results here immediately in sigmaGuide because your procedure is then intransparent for all participants. It makes sense here to go through the steps of the process with the participants of the workshop and proceed according to the instructions in sigmaGuide. Thus, note each process step and each negative influence on sticky notes, e.g., on a Post-It. Photos of activities, methods, resources, intermediate outputs, and negative influences are useful if the process is documented after the inspection on a flip chart.

I always use sticky notes, which I can move and change if necessary. In this way, all participants can gradually follow, comment on, correct, and cooperate with the structure of the process mapping and its analysis. Only if all workshop participants agree with the result, you should transfer the important results into sigmaGuide. And here in the Project-Story-Book the Tool-Frame is the placeholder for photos of the workshop with the participants and the Process-Mapping-Analysis flip chart.

The same applies to the second workshop on Root-Cause-Analysis. Please build the Hierarchy-Tree together with competent workshop participants and check and correct it until all participants agree. Mark the anchor points for solution ideas and collect solution ideas from the participants.

Finally, please photograph the result again and insert it into the Project-Story-Book, if possible, with a photo of the workshop participants.

The C&E Matrix also has a special status. The matrix itself and the derived C&E Heatmap from the sigmaGuide are not well suited for a presentation due to their size. Of course, you should include a copy of both tools. For a better understanding please additionally summarize the most important relationships in an overview matrix (Fig. 5.2).

In this template I have contrasted the most important influences of input, methods, and resources in the columns with the most important problems in the rows. And I transformed the respective risk values from the C&E Heatmap into the

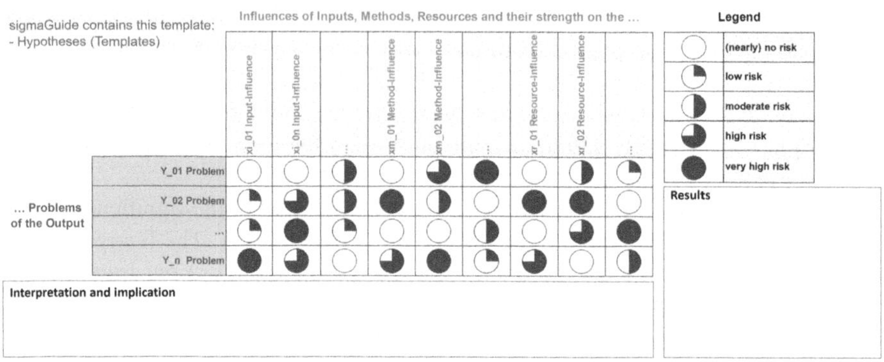

Fig. 5.2 Project-Story-Book: Summary of the C&E Matrix/Heatmap with important influence (*x*) problem (*Y*) relationships and the risks that the influences trigger or increase the problems, represented by 5-level Harvey Balls

Define-Steering				
Master-Black-Belt	**Proceed to next-Phase**	**Remarks**	**Date**	**Contact/ Verification-ID**
Dr. Reiner Hutwelker	yes	to be filled in by the Master Black Belt	X.X.XXXX	reiner.hutwelker@sigmalogic.de
	no			
Sponsor	**Proceed to next-Phase**	**Remarks**	**Date**	**Contact**
	yes	to be filled in by the Sponsor (if you like then use the Sponsor-Checklist in sigmaGuide)	X.X.XXXX	
	no			

Fig. 5.3 Project-Story-Book: decisions of the Sponsor and Master Black Belt in the DMAIC steerings on necessary corrections or on the continuation with the next phase

five levels of the descriptive Harvey Balls. This format makes it easier for everyone to understand the core message of the MEASURE phase and to develop an understanding of the derived hypotheses.

At the end of each phase there is a slide showing the results of the phase steering. Here the Master Black Belt can decide whether the phase has been successfully completed. The second decision of the Sponsor is decisive for the transition to the next phase. *No* means: Rework. *2 x Yes* means: *Go to the next phase* (Fig. 5.3).

All right? 2 × Yes? Then let us go to the next tool!

Tasks

Business-Projects: Please prepare your Project-Story-Book up to the Project-Definition.

Standard-Project: Please prepare your Project-Story-Book up to the Project-Definition.

To illustrate the problem of your hometown you must take some photos of garbage places around trash-cans. Please do this using the *World Cleanup App* (https://www.worldcleanupday.org/map-it). With this app you can additionally publish the geodata of the photographed places into your Project-Story-Book.

You probably do not have a Sponsor for your project yet, but maybe you will find a manager from your garbage disposal service later who is willing to be your Voice of Business. Perhaps the photos and their geodata from your Project-Story-Book can convince this manager to support you implementing your project.

DEFINE: Part 3

6

6.1 SIPOC[1]

The SIPOC is the first tool of the DMAIC for implementing a project. The columns of this simple list are predefined for specifying (internal) suppliers (S), their inputs (I), the related process steps (P), the resulting (intermediate) outputs (O), and their customer (C). In my opinion, the SIPOC is rather worthless in itself. But it fulfills important purposes in the tool chain.

The first purpose is to determine the significant (intermediate) outputs, which then can be evaluated by the customer and manager in the subsequent VoC and VoB to identify the critical problems (CtQ).

The second purpose is to define the limits of the project from the last relevant output to the first relevant input and thus to define the field of the project.

And the third purpose is the transparency of the core process steps within this project field.

The limiting output should be the output with the chronological last problem from the problem statement in the Project-Definition. The video: *Cookie du Chef* (Sect. 2.3) showed that our customers did not like the delivered cookies and the delivery date was too early.

During the process steps, the cookies reach higher development stages. These stages can be specified by the modified or added attributes of the cookies. I always specify the output at least by the last attribute added. The output in the last step was the delivered cookie and I enter: *Cookie (delivered)* and thus name the output and put its latest attribute in (brackets) behind it. This output is the lower limit of our SIPOC, and I enter it in the lower right cell (Fig. 6.1).

The first input should be the first possible trigger of the output problem. This limits the project field to the relevant process steps, which include all relevant outputs, their problems, and their possible triggers.

[1]Video 16 (https://www.youtube.com/channel/UCShsy50l5dFsoSXU__hBKSA).

© Springer Nature Switzerland AG 2019

R. Hutwelker, *Six Sigma Green Belt Certification Project*, Management for Professionals, https://doi.org/10.1007/978-3-030-31915-1_6

SIPOC					
Process-Step	Supplier	Input (xI)	Process (xMR)	Output (Y)	Customer
1	Customer (Child)	Request (cookie-type & delivery-date)	check Disposability	Order (cookie-type; delivery-date)	Logistic & Procurement Service
2		Order (cookie-type; delivery-date)	determine necessary & missing Ingredients	Shopping-List	Logistic & Procurement Service
3	Supermarket	Ingredients	assemble Ingredients	Ingredients (complete)	Logistic & Procurement Service
4		Start-Signal	prepare Workspace	Workspace (clean)	Production Support
5		Ingredients (complete)	weigh Ingredients	Ingredients (weighed)	Production Support
6		Ingredients (weighed)	knead Ingredients	Dough (ball)	Chef du Cookie
7		Dough (ball)	roll out Dough	Dough (sheet)	Chef du Cookie
8		Dough (sheet)	cut out Cookies	Cookies (raw)	Chef du Cookie
9		Cookies (raw)	bake Cookies	Cookies (baked)	Chef du Cookie
10	Tin Factory	Tin	pack Cookies in Tin	Cookies (boxed)	Production Support
11		Cookies (boxed)	deliver Cookies	Cookies (delivered)	Logistic & Procurement Service
12					

Fig. 6.1 SIPOC (Supplier–Input–Process–Output–Customer): Specify the project limits by the output with the last problem and the first influential input and structure the process-field row by row by the important process steps leading to the important outputs for Voice of Customer/Business

Fig. 6.2 Practice tips for the work with SIPOC

Practice Tips:

- Indicate the subsequent development stages of the objects in Input and Output by adding the respective latest characteristic attribute of the object in brackets, e.g.: Cookies (baked).
- Typical number of (core) Process-Steps in a SIPOC: 8 – 12
- To shift an entry: never use cut & paste – instead use: copy & paste and delete unnecessary cells
- If an Output in one step serves as an Input in a subsequent step, then copy the Output and paste it as new Input
- If more than one object enters the process as Input, then only enter the new external object as Input
- Describe the Activities of the Process with a verb, followed by a noun, e.g.: bake Cookies

Since I do not know if the customer did not like the cookies, because (a) she actually requested a different type of cookie or because (b) the cookies just tasted bad, I enter the request for cookies with their attributes in the upper right cell: *Request (cookie-type and delivery-date).*

Now we know the field of the project, from the first input on the top left side to the last output on bottom right side. The next task is to break down the enclosed processes of this field into core process steps line by line in the rows of the SIPOC (Fig. 6.2).

Here it is important to find the balance, not to become too detailed as in the later Process-Mapping and not to sketch too roughly. sigmaGuide can take up a maximum of 16 core process steps; typical in practice, however, are 8–12 entries.

In the first process step, the *request* is *checked* and transformed into an: *Order.* This *order* has the same attributes as the *request*, and I put it in brackets here as well: *Order (cookie-type and delivery-date).* This is our first core process step: Input–Process–Output.

Now I determine all intermediate outputs and define their relevance by the fact that they are either significant new outputs, or attributes of a preceding output have been transformed significantly.

Thereafter, the next output is a *Shopping-List*, then we have the *Ingredients (completed)*, the *Workspace (clean)*, the *Ingredients (weighed)*, we have the

Dough (as a ball) and the *Dough (as a sheet)*, transformed to the *Cookies (raw)*, the *Cookies (baked)* and *(boxed)*. Then comes the last output *Cookies (delivered)*.

To close the gap in the table, I copy this entry, paste it under *cookies (boxed)*, and delete the last entry. Please never use: Cut and Paste in sigmaGuide, otherwise the cell references in the following tools will be wrong.

Next, the inputs are specified. We know already the *request* for cookies. In a consecutive process chain, the output of an upstream process step becomes the input of the downstream step. Here you can see that the next input—the *order*—is the output of the preceding step.

If an output in one step serves as an input in a following step, please always copy the corresponding output and paste it as new input. In the following tools, inputs and outputs are analyzed separately. Only if identical inputs and outputs match exactly in their spelling, sigmaGuide can make this separation. In the Input–Analysis, for example, the outputs are also listed, but are enclosed in [squared brackets]. This means that these inputs are also available as outputs and have already been evaluated as output in the preceding tool: Critical-to-Quality (VoC/VoB; CCR/CBR; CtQ). And that is the way it is supposed to be.

A new input is the *ingredients* as well as the time *signal* to start baking. In the next steps we again can copy the outputs as subsequent inputs, the *ingredients (complete)*, *(weighed)*, the *dough (as a ball)*, *(as a sheet)* and the *(raw) cookies*. A new input is the *tin* in the next step. And this cell should actually contain: *Tin* and *cookies (baked)*. This however would give us two entries later in the Input–Analysis, of which only the new input—the *tins*—are relevant. Because the *cookies (baked)* will be evaluated in the tool: Voice-to-Critical. To avoid this redundancy, I will leave it at the new input. *Tin*. The last input is again a previous output: *Cookies (boxed)*.

Now let us enter the activities of the process steps. Describe each process step with a verb and a noun to illustrate the transformation of the input into the output. You probably remember the Process-Problem-Solving Model (Sects. 2.4.2 and 2.4.3): Inputs and outputs are always tangible or intangible objects, such as a cookie or date. Process steps are always activities, such as *baking* or *entering*.

To transform the request into the order it is necessary to: *Check the disposability of cookie-type and delivery-date*. To transform a *dough (ball)* into a *dough (sheet)* it is necessary to: *Roll out the dough*.

At least the *suppliers* of the inputs are entered and the *customers* of the outputs. Now we have divided the entire project field into the core process steps and can very easily track the development stages of the cookies based on their (intermediate) outputs and attributes.

Please check the consistency of all core process steps. If your SIPOC now matches reality, then you can copy it into your Project-Story-Book.

Questions (SC = Single Choice; MC= Multiple Choice)

Video 16/ Ch. 6.1/ #1 (MC)	What are the four most important purposes of the SIPOC?	
1	Determine the important (intermediate) outputs	
2	Define the limits of the project field	
2	Inspect the used methods and resources of the process steps	
3	Get an overview of the influences of the process	
4	Evaluate the outputs of the process steps	
5	Identify the most important activities of the process steps	
6	Make the core process steps transparent within the project field	
7	Identify the interviewpartner for the VoC	
7	points	

Video 16/ Ch. 6.1/ #2 (SC)	Which word types should be used to specify inputs, process steps and outputs?		
		noun	verb + noun
1	Inputs		
2	Process-Steps		
3	Outputs		
3	points		

Video 16/ Ch. 6.1/ #3 (MC)	What should be additionally specified in brackets for each output if possible?	
1	The number of variants of a product/ service	
2	The problem of the output	
3	The last added/modified attribute of the output	
4	The achieved development stage of the output	
4	points	

Tasks

Business-Projects: Please prepare the SIPOC for your project.

Standard-Project: For your orientation some entries of the SIPOC are already prefilled. You can customize these entries according to your needs. And you should enter some additional information to complete your SIPOC. It will probably not be very complex, because 5–8 core process steps can be enough to adequately describe the project field of this Standard-Project.

The already entered Supplier (occasional-litter and heavy-litterer) refer to personality types with different littering behavior due to underlying extrinsic and intrinsic motivation. Their different motivation determined the different entries in the core process steps of SIPOC to distinguish the *prudent* from the *carefree* littering (Table 6.1).

Table 6.1 Types of garbage litterer, their behavior, and underlying extrinsic or intrinsic motivation (adapted from: IGSU, IG saubere Umwelt, Zurich, Switzerland https://www.igsu.ch/de/littering/littering-typen/)

extrinsic motivated	**3. Cautious-Avoider:** Concerned with the public image, feels observed, doesn't want to be caught	extrinsic motivated	**3. Team-Dropper:** Can be instigated by others, does not want to make himself unpopular/ laughable	extrinsic motivated	**3. Hard-Core-Dropper:** Littering is fun, arises from the wish to provoke others and to demonstrate own strength/independence.
intrinsic motivated	**2. Green-Mind:** Avoids littering from personal conviction, disposes of garbage at home if necessary	extrinsic motivated	**2. Smart-Dropper:** Finds a hidden place for disposal or throws own garbage on the ground to already lying garbage.	extrinsic motivated	**2. Purposive-Dropper:** Has no reason to dispose waste correctly, because that is the assigned task of the garbage removal servic and secures their jobs
intrinsic motivated	**1. Green-Leader:** Correct disposal of all gargabe types, from personal conviction, sometimes picks up garbage from others, tries to convince others	intrinsic motivated	**1. Stress-Dropper:** Has no time to dispose of the waste correctly	intrinsic motivated	**1. Life-Style-Dropper:** Littering is part of life, waste is deposited where it arises, in public places and at home, no matter if somebody disposes of it or not
	1. Non-Litterer		**2. Occasional-Litterer**		**3. Heavy-Litterer**

A related questionnaire is presented in the chapter: Data-Collection-Plan (Sect. 7.5.2).

6.2 Voice-to-Critical[2]

sigmaGuide combines the steps for determining the: Voice of Customer and Business (VoC/VoB), the Critical Requirements of Customer and Business (CCR/CBR) and the Critical to Quality (CtQ) in one tool: Voice-to-Criticals.

6.2.1 Procedure

To this end, the voices of customers and managers on their wishes and complaints on each of the defined outputs of SIPOC are first collected (VoC/VoB). The attribute of the output concerned is derived from these wishes and complaints (Fig. 6.3).

In the first case, I would like to point out that increasing transparency is one of the basic tasks of the project. In the second case, I ask managers not to make their proposed solutions binding in this DEFINE phase, but to postpone them until the IMPROVE phase. This, however, does not apply to the immediate implementation of *quick wins*, if their success is guaranteed.

Based on this attribute the critical requirements are defined (Critical Customer Requirements: CCR and Critical Business Requirements: CBR) and the deviations from these requirements are defined as problems (Sects. 2.4.2 and 2.4.3). Then the type of problem is determined, as a quality, availability, or consumption problem. According to the needs of the customer the problems are then categorized according

[2]Video 17 (https://www.youtube.com/channel/UCShsy50l5dFsoSXU__hBKSA).

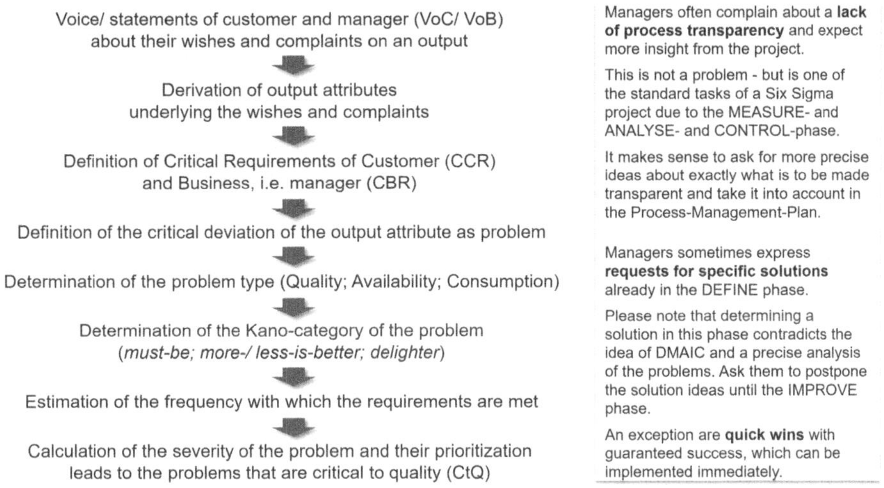

Fig. 6.3 Voice-to-Critical leads from the Voice of the Customer (VoC)/Voice of Business (VoB) via the Critical Requirements of Customers and Business (CCR/CBR) to the critical problems (CtQ)

to the Kano-model. Afterwards, the customer estimates the frequency with which his or her expectations were fulfilled in the past.

Based on this frequency and the Kano-category of the attribute, the severity of each problem is calculated. Then the problems are prioritized to identify the problems, which are Critical to Quality (CtQs). The identification of these CTQs is the purpose of this tool: Voice-to-Criticals.

The necessary information is preferably collected in interviews with the involved customers and managers. Sometimes we pretend to already know the answers of customers and managers without actually questioning them. But this is often a misbelief. No commissioner would be content to solve a case himself in the office without questioning the witnesses. And a Six Sigma project has a certain resemblance to a criminal case. But of course, we are not dealing with murder, we are dealing with problems. But having an attitude like Sherlock Holmes cannot hurt.

6.2.2 Interviews

So, get on your way to the customers of the (intermediate) outputs and the responsible managers of the process steps from the SIPOC—if possible. This can also increase the acceptance of the project in your work environment because (internal) customers are typically rarely interviewed on their needs and problems in practice.

If you capture these voices, you will get a lot of information and wishes. However, it is not always a matter of requirements and sometimes it is difficult to distinguish the requirements from further statements. Many process owners and Sponsors additionally wish for greater process transparency and sometimes express wishes as to what a solution to the problems could look like.

Procedure to collect the necessary information, e.g. in interviews:

1. Select a relevant source and collect answers at least to the following questions:
2. Which of these SIPOC outputs is causing you trouble?
3. What do you require from this output?
4. What complaints do you have about these outputs?
5. Which quality of output does your request and complaint relate to?
6. How should this property be pronounced in the best case?
7. How is this characteristic pronounced in the worst cases?
8. What is their satisfaction if your expectations are fully met?
9. What is their satisfaction if your expectations are not met?
10. What percentage of the outputs received meets your requirements?

Sources to collect the Voice of Customer (VoC):

1. Internal/ external customer
2. Internal quality control,
3. Complaints (sales/ customer care),
4. Warranty cases,
5. Dealers,
6. Customer satisfaction surveys,
7. Social media and
8. Own experiences and experiences of friends/ colleagues

Sources to collect the Voice of Business (VoB):

- Personal interviews with:
  - Process owner
  - Sponsor (if not the same)

Fig. 6.4 Procedure of the Voice-to-Criticals. Use as many information sources that are available to you—interviews are preferred

The information you need to collect on each SIPOC output is derived from the answers to the questions in this list (see Figs. 6.4 and 6.5). I will introduce them as soon as I enter the data in the tool.

In many companies, however, it is forbidden to interview external customer. But there are additional sources to also obtain Voices of these Customer(s), like voices from internal quality control, sales, customer care, warranty cases, dealers, customer satisfaction surveys, social media, your own experiences and experiences of friends and colleagues. In the most complex case I know, thousands of voice recordings from customer care were analyzed automatically.

After you have recorded the voices from all relevant sources, i.e., the answers to the questions, you can enter them into the tool.

6.2.3 Transferring Answers to the Tool

Voice-to-Criticals again leads you with a controlled dialogue through its 10 questions for each relevant output to translate the collected voices into evaluated problems.

I want to enter the complaint of a customer about the taste of our cookies first and therefore also select: *Customer* from the dropdown list (see Fig. 6.5).

In the next question to the affected output I enter: Cookie*s (baked)*. All outputs of the SIPOC are listed here. The output state: Cookie*s (baked)* is appropriate here because the taste does probably neither changed by *packaging* nor by *delivery*.

In the video: Cookie *du Chef* (Sect. 2.3) you could see the request in the customer's letter. It says that the cookies should "taste like grandma's." This is the demand in the customer's voice. And in the interview the customer initially complained that: "Cookies taste of nothing."

Y_01		
Whose Requirement do you want to specify?	Customer	Source of Requirement: Customer
	Please select an answer.	
Which Output does the the Customer want to evaluate?	**Cookies (baked)**	Output: Cookies (baked)
	Please select an answer.	
What does the Customer require from COOKIES (BAKED)? COOKIES (BAKED) should ...	**Cookies taste like those from Grandma**	Voice of Customer (VoC): Demand
	Please quote the statement of the Customer.	
Which deviation of COOKIES (BAKED) from the Requirement is problematic for the Customer ?	**Cookies taste of nothing**	Voice of Customer (VoC): Complaint
	Please quote the statement of the Customer.	
Which attribute of COOKIES (BAKED) is addressed in this Demand and Complaint?	**taste**	Attribute of COOKIES (BAKED): TASTE
	Please enter just one Noun.	
How should TASTE be according to the Requirement of the Customer?	**mellow-sweet**	Critical Customer Requirement (CCR): COOKIES (BAKED) TASTE MELLOW-SWEET
	Please enter just one adjective.	
Which deviation of TASTE is problematic?	**crumbly-bland**	Problem: COOKIES (BAKED) TASTE CRUMBLY-BLAND
	Please enter just one adjective.	
To which category does the Requirement TASTE MELLOW-SWEET vs. CRUMBLY-BLAND belong?	**Quality (Faultlessness/ Fulfilment of Purpose)**	Requirement-Category of TASTE: Quality
	Please select an answer.	
Which relevance does the Requirement COOKIES (BAKED) TASTE MELLOW-SWEET have according to the Kano-Model?	**Must-Be**	Kano-Category of TASTE: Must-Be
	Please select an answer.	
To what extend is the Requirement on TASTE of COOKIES (BAKED) actually fulfilled?	**20%**	Requirement for TASTE of COOKIES (BAKED) is fulfilled to: 20%
	Please enter a value between: 0% - 100%.	
Y_01: Output: Cookies (baked) | Voice of Customer (VoC): Demand: Cookies taste like those from Grandma; Complaint: Cookies taste of nothing | Critical Customer Requirement (CCR): COOKIES (BAKED) TASTE MELLOW-SWEET | Problem: COOKIES (BAKED) TASTE CRUMBLY-BLAND | Requirement-Category of TASTE: Quality | Kano-Category of TASTE: Must-Be | Requirement for TASTE of COOKIES (BAKED) is fulfilled to: 20% | The Severity of being unsatisfied with TASTE of COOKIES (BAKED) is: 92% (Rank 1/ 5).		

Fig. 6.5 Voice-to-Criticals: Please answer the questions and check the summary. If the statements in the summary are pointless or cumbersome, please adapt the answers

The next step is to filter out the attribute of the output that underlies the demand vs. the complaint. This is not always easy with quality attributes, in contrast to the attribute of availability and consumption. Therefore, I contrast the demand and the complaint in this tool to simplify the identification of the underlying attribute. You may not always be able to determine quality attributes right away, especially for intangible outputs. But with a little practice, you will find it easier and easier.

In the example it is easy to derive the affected attribute of our cookies as their: "taste." On my question, how the "taste" should be, the customers answered: "mellow-crunchy-sweet."

Additionally, the customer specified their complaint about our cookies in: "cookie taste crumbly-bland." And this is the problem.

As already mentioned and easy to understand, the requirement on the taste and thus also the problem belongs to the category of quality.

Question 9 asks for the relevance of this requirement according to the Kano-Model. The customer consistently told me that they expect a "mellow-crunchy-sweet taste" as a prerequisite to buy cookies and that they get angry if they taste "taste crumbly-bland." Thus, I select: "must-be" as the appropriate category of the Kano-Model.

6.2.4 Kano-Model

You probably already know the Kano-Model, but I will introduce it again for some adaptions I made to integrate it into a continuous tool chain.

Noriaki Kano distinguishes three types of requirements for a product or service according to the relationship between the fulfillment of the requirement and the resulting satisfaction. These three requirement types are reflected in a two-dimensional model as surfaces below and above the characteristics of these types.

Every must-be is a basic requirement. Its fulfillment is implicitly expected as the norm and leads to dissatisfaction if it is not fulfilled. Must-bes vary between indifferent or moderately satisfied if the attribute completely meets the requirement. And the customer is angry if the requirements are completely missed. Examples of these basic requirements for cookies are that they must-be mellow, crunchy, and sweet. If cookies taste crumbly-bland or are toxic, we probably would not buy them.

The **more-/less-is-better** performance requirements can lead to dissatisfaction if they are not met and to excitement if they are very pronounced. Our cookies should be fresh (the fresher, the better), inexpensive (the cheaper, the better), and low carb (the lower, the better). We are probably enthusiastic when they are almost free and would not buy them if they are too expensive. The fulfillment of more-/less-is-better requirements is typically linear related to satisfaction.

Delighters belong to the marketing tools. They can make the difference in competition if products or services from competitors are similar in fulfilling the must-bes and more-/less-is-better requirements. Our customers would probably be thrilled if cookies contribute to physical fitness, are harmless to the teeth, and are packed in environmentally friendly boxes (Fig. 6.6).

Delighters assure excitement. They are irrelevant for Six Sigma projects in that they cannot compensate for deviations in the must-bes or low more-/less-is-better levels. We do not buy cookies that are harmless to teeth but taste bad.

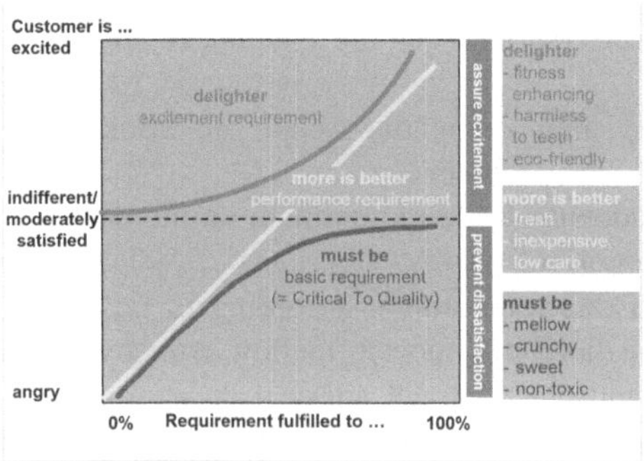

The **KANO-Model** serves to:

- Classify the requirements of the Customer (CCR) and Business (CBR) and
- Identify the critical requirements (CTQ´s= Critical to Quality), i.e. severity of the output problems.

There are three categories, specified by their characteristics:

1. **must be:** These are the basic requirements of the customer/ business which have to be fulfilled to prevent dissatisfaction.
2. **more/ less is better:** These are the performance requirements of the customer/ business which should be fulfilled either as high/ as low as possible or straight to a certain value within a specific range.
3. **delighter**: These are the excitement requirements, important for marketing but unimportant for Six Sigma, because they are not necessarily expect, but lead to excitement if given.

Focus is on the **must bes**. These attributes determine the CTQ´s of your product/ service, because they cannot be compensated by other attributes of your product/ service.

Fig. 6.6 Kano-Model: Identify the critical requirements of the customer and the business (CtQs) from the severity of the *must-be* problems of the outputs

The **must-bes** prevent from dissatisfaction but do not create excitement. They are the important requirements in Six Sigma. If must-be attributes are not met as expected, then they are probably critical to quality (CtQ).

The **more-/less-is-better** attributes of our products or services may also be relevant for Six Sigma if they can cause dissatisfaction. It is a decision of your Sponsor, expressed in the objectives of the Project-Charter, which attributes of your product or service you should include in your project.

6.2.5 Transferring Answers to the Tool

After determining the type of requirement according to Kano, the customer or manager respectively must estimate in the last question the extent or frequency to which the requirement is fulfilled as expected. Our customers told me that in the past 20% of cookies met their requirements on: *Taste.*

In the Feedback at the bottom of each block all given answers are summarized (see Fig. 6.5). Please read these summaries after your data entry and correct entries if there are duplicates or blurs. On the bottom of this summary you can see the entry: *The Severity of being unsatisfied with the TASTE of COOKIES (BAKED) is: 92%.* And this is the 1st rank of 5.

The calculated severity results from an exponential function in which the Kano type, in this case a *must-be*, is combined with the fulfillment level: 20%. The ranking: First of five (1/5) results from the fact that I have already entered the other four problems regarding availability and consumption.

And the next problem relates to the availability of: Cookies (delivered). Here the customer demands that the cookies are delivered on the order date and complains that they were actually delivered too early. Requirements on availability make it easy to derive the attribute of the output, because it is usually a question of the required quantity on the required date. According to the customer, the requirements are met if the cookies are delivered a maximum of ± 1 hour—too early or too late. In this case, the exact specification limits that will later become important in the Data-Collection-Plan and in the Process-Management-Plan can already be determined here. *Being on time* is a typical more/less-is-better requirement, just like: *the faster the better.*

Finally, the customer estimated that her requirement for punctuality of this output was met in 40% of cases.

A summary is then again created from this information. The problem: Cookie*s (delivered)* delivery more than 1 hour too early/late has a calculated severity of 60% and takes the second place of five problems (2/5).

Scrolling down the list in the example file of sigmaGuide you will find the definition of further problems of resource consumption.

6.2.6 Results

The next worksheet in sigmaGuide contains two tables with results. One includes the detailed summaries for each block that I have just presented. The second table includes only the most important information: Voice of Customer or Business, the critical requirements (CCR/CCB) and problems, their respective Kano category, the severity and the ranking of each problem in comparison to all defined problems. You can sort this table by: severity with: *Ctrl* + *Shift* + *S*. Then your critical problems—the CtQs—will be at the top of the list (Fig. 6.7).

Finally, sigmaGuide creates two charts for the Voice-to-Criticals. The Chart Y CTQ bar chart shows the severity of the problems compared in a bar chart (Fig. 6.8).

Let me explain the second chart in more detail. It shows the characteristics of the three Kano types, with the problems located at their specific positions. But as you can see, I have mirrored the characteristics of the *must-bes* and *delighter*. There is a reason for this:

If you look at the original curve of the *must-be* characteristic, you can see that a *must-be* that is fulfilled in 50% of the cases leads to an almost: *indifferent/moderately satisfied*. Now imagine, e.g., that in 50% of the cases your telephone calls will be cancelled or 50% of your travels will be interrupted due to a breakdown. Of course, past experience also puts into perspective how satisfied we are with a 50% performance, but many people would probably say that even with this unreliability they are more annoyed than being: *indifferent*. To take this into account, I mirrored the *must-be* characteristic on its axis. I also mirrored the *delighter* characteristic. Imagine, after 50% of your phone calls or trips you receive a free ticket for an event of your choice. This is unusual for communication and travel companies and therefore, these

Summary: Voice of Customer (VoC), Voice of Business (VoB), Critical Requirements (CCR/ CBR), Problems, Severity, KANO and CtQ-Rank							
Y	Voice	of ...	Critical Business Requirement (CBR) or Critical Customer Requirement (CCR)	Problem	Kano-Category	Severity	Critical to Quality (CtQ) Rank
Y_01	Cookies (baked) Cookies taste of nothing	Customer	CCR: Cookies (baked) taste mellow-sweet	Cookies (baked) taste crumbly-bland	Must-Be	92%	1
Y_02	Cookies (delivered) delivered too early	Customer	CCR: Cookies (delivered) delivery in time	[illegible] delivery > 1 hour too early/ late	[illegible]	60%	2
Y_03	Cookies (boxed) waste of energy	Management	CBR: Cookies (boxed) energy-consumption minimal	Cookies (boxed) energy-consumption > 10% waste	More/Less-is-Better	30%	3
Y_05	Cookies (baked) Cookies are too big	Management	CBR: Cookies (baked) diameter > 9cm and < 10 cm	Cookies (baked) diameter > 10 cm	More/Less-is-Better	30%	3
Y_04	Cookies (boxed) waste of ingredients	Management	CBR: Cookies (boxed) ingredient-consumption minimal	Cookies (boxed) ingredient-consumption > 10% waste	More/Less-is-Better	20%	5

Fig. 6.7 Voice-to-Critical: Summary and details of VoC/VoB, Critical Requirements, CTQs for all Ys. This information about Y is the anchor of your project

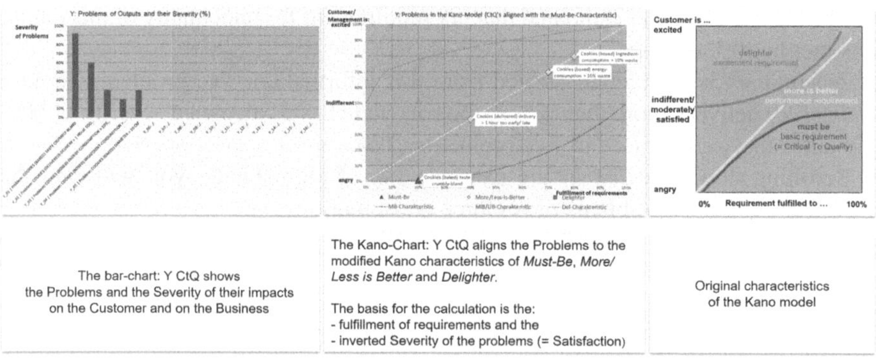

Fig. 6.8 Voice-to-Critical: Bar-Chart with the Severity of the Problems (Y) and adapted Kano-Chart with the fulfillment of requirements and calculated satisfaction

bonuses are *delighter*. And I would be almost completely enthusiastic about the 50% fulfillment of these bonuses.

Therefore, I am convinced that the mirrored characteristics of the *must-bes* and *delighter* reflect reality better than the original ones. The *more/less-is-better* characteristic I left as in the original model.

Now you are ready for the interviews. They are obligatory for the Green Belt certificate. But itis worth it. Be careful that you get not lost in information and make sure you get all the answers you need. Stay persistent and ask as long as you are not sure that you have understood the answers.

Questions (SC = Single Choice; MC = Multiple Choice)

Video 17/ Ch. 6.2/ #1 (SC)	**Please bring the following steps to completing the tool: Voice-to-Criticals in their correct order.**					
		1st step	2nd step	3rd step	4th step	5th step
1	Define Critical Customer/ Business Requirements (CCR/ CBR) on the attributes					
2	Determine the problem type as a quality, availability or consumption problem					
3	Identify wishes and complaints in the Voice of Customer/ Business (VoC/ VoB)					
4	Define the critical deviation of an attribute as a problem of the output					
5	Derive important attributes of the output from the wishes and complaints					
5	points					

Video 17/ Ch. 6.2/ #2 (SC)	**Please assign the appropriate Kano category to the reactions of a customer resulting from the fulfillment and non-fulfillment of the requirements.**				
		angry if not fullfilled	indifferent if not fulfilled	indifferent if fulfilled	excited if fulfilled
1	must-be				
2	more-/ less -is-better				
3	delighter				
3	points				

Video 17/ Ch. 6.2/ #3 (MC)	What information is used to determine the critical problems, the CtQ's?	
1	CCR/ CBR	
2	Kano-category (must-be, more-/ less-is-better, delighter)	
3	VoC/ VoB	
4	Extent to which requirements are met	
5	Problem type (quality, availability or consumption)	
6	Calculation of the severity of the problems	
7	Prioritization of the problems based on their severity	
7	points	

Tasks

Business-Projects: Please prepare the Voice-to-Critical for your project.

Standard-Project: For your orientation two problems from the Project-Definition are already specified as Y01 and Y02 from the perspective of the garbage disposal service (Voice of Business). You can customize the entries according to your needs. Additionally you should specify at least one more problem (Y03), derived from the third problem of the Project-Definition. Here citizens as customer complain (Voice of Customer) that the garbage is not completely removed from streets/places by the garbage disposal service. In the tool Voice-to-Criticals you will find more problems and specify them from the perspective of the business or the customer if you really interview responsible managers of the garbage disposal service and some citizens.

6.3 Project-Charter[3]

The Project-Charter does not belong to the problem-solving tools of the DMAIC but supports to organize and control the project. It serves as a contract between the Green or Black Belt and the Sponsor (Fig. 6.9).

The Project-Charter summarizes some important already collected information and supplements a project-title, the problems, scope, team, targets, and deadlines. The given information is divided into the fields: Business-case, Voice of Customer, Voice of Business, process and output, and the first solution-ideas.

First the project gets a *name* that reflects its purpose.

The: *Business-case* summarizes information from the Project-Topic and the Project-Definition. Some companies require a more detailed description of the strategic importance of the project. If so, you should ask your Sponsor what further

[3]Video 18 (https://www.youtube.com/channel/UCShsy50l5dFsoSXU__hBKSA).

Project-Charter	**Project-Name** Improve the performance of Cookie production
Business-Case The Product COOKIE is a tangible final Output for external Customers and is in the Creation Process BAKE COOKIES within a year 13 - 52 times generated. Important Input of the Process to generate the Product COOKIE is: BUTTER, SUGAR, FLOUR, CHOCOLATE.	**Process & Output** Product/ Service: Cookie Process: bake cookies
Voice of Customer (VoC) The satisfaction of the external customers with the: - Quality of COOKIE is: 20%. - Availability of COOKIE is: 40%.	**Problems** Y_01 | Cookies (baked) taste crumbly-bland Y_02 | Cookies (delivered) delivery > 1 hour too early/ late Y_03 | Cookies (boxed) energy-consumption > 10% waste
Voice of Business (VoB) The satisfaction of the process-owners with the Consumption in the Creation Process of the COOKIE is: 40%. The total costs of the specified 3 problems are estimated by 100€ / year. They are primarily the result of quality costs due to scrap and additional expenditure. The solution of the problems is rated as: - major URGENT (80%-Level) / - major IMPORTANT (80%-Level)	**Solution-Ideas** more sugar in the mix
Comment	**Comment**

In Scope	Out of Scope	Management			
in: Vanilla Cookies	out: Chips	Sponsor	Mr. B. Aking	Supplier	Supermarket
in: Chocolate Cookies	out: Bread	A_ccountable	Chef du Cookie	Customer	Charlotte
in:	out:	Controlling	Mrs M. Oney	Customer	Lili
in:	out:	...?		...?	

Targets	Timeline	Experts			
Y_01 Improve taste of Cookies (grade of 1,5 in customer rating)	30 October 20xx	Black-Belt	Mr B. B. Elt	Master-Black-Belt	Reiner
Y_02 Deliver ordered Cookies on time (+/- 1h)	30 October 20xx	Green-Belt	Y. Ou	...?	
Y_03 Reduce consumption of energy (- 20%)	30 October 20xx	Expert	Mrs D. Ough	...?	
Y_04 Reduce consumption of ingredients (- 20%)	30 October 20xx	Expert	Mr E. Quipment	...?	

Timeline	Define	Measure	Analyse	Improve	Control*	Control	End
Target-Date:	03.08.20xx	31.08.20xx	28.09.20xx	26.10.20xx	23.11.20xx	01.06.20xy	28.06.20xy
Completion-Date:							
Evaluation:	days remaining : 58	days remaining : 86	days remaining : 114	days remaining : 142	days remaining : 170	days remaining : 360	days remaining : 387

Fig. 6.9 Project-Charter: The contract for your project with business-base, VoC/VoB and solution ideas. Please select the problems and add scope, team members' targets and deadlines

information he or she expects. And you could also append this extended business-case in your Project-Story-Book.

The field: *Problems* contains three list boxes with the problems from the tool: Voice-to-Critical. Here I choose the three CtQs with the highest severity (see: Summary in the Voice-to-Criticals). And this is the preferred approach. However, your Sponsor may want to set other priorities. Thus, discuss with your Sponsor which three problems should be dealt with in the project.

Additionally, it is important to define the *Scope* of the project. You may include or exclude certain outputs or process steps to further sharpen the content of the project. Then all involved persons can be certain what they can or cannot expect from the project.

In our bakery the Sponsor and the Belt agreed that vanilla and chocolate cookies are in the scope, bread and chips are out of scope.

This expectation is further specified in the *Targets*. Targets are often formulated as fuzzy declarations of intent. And you will find terms like: Optimize or maximize. But this is more of an activity than a state and there can be different opinions about what is optimal or maximum. Thus, a target always specifies a state that is to be achieved.

The *smart*-criteria can help to distinguish between suitable and unsuitable targets. The letters of this acronym stand for:

s: means **specific**: It must be clear what exactly the target state is.

m: means **measurable**: A measure and a target value or value corridor should indicate the success criterion.

a: means **attractive**: It is difficult for people to stand up for goals with which they do not identify. In any case, it is important that there is an incentive to commit to the target. In our case, the prospect of the Green Belt certificate is probably sufficient. However, if a Belt is to implement further projects later, as usual in addition to the regular working hours, then an incentive for the Belt and its team can make the target additionally attractive and motivate rapid implementation.

r: means **realistic**: Realistic goals are demanding but are also subject to specific conditions and feasibility. Targets that demand 100% performance are unrealistic, even if they must be demanded in the area of work safety. But even the extreme challenging goal of 6 sigma still allows for defects (3.4 ppm).

The consequences of unrealistic targets are that they lead to fear or frustration if they are set too high and too low targets lead to boredom. Achievement-motivated humans set their goals always only a little higher than their so far reached results.

t: means **terminated**: And a deadline for the target is important to speed up and control the project implementation.

Our targets are based on the problems. In the Project-Charter we do not have to invent anything new. We simply turn the selected problems upside down.

For this purpose, the numbers of the three selected problems are already preset for the targets. Conform to the smart criteria I specified:

- Y_01: Improve taste of cookies (grade of 1, 5 in customer rating),
- Y_02: Deliver ordered cookies on time (±1 hour),
- Y_03: Reduce consumption of energy (−20%) and
- Y_04: Reduce consumption of ingredients (−20%).

All targets should be achieved until the end of October.

For the fourth target here, there is no corresponding problem listed in the problem field, as there are only three problems listed. But it is possible to manually enter any additional target that seems necessary for the Sponsor. But take care, (a) that it also fits to the smart-criteria and (b) that your project does not become too complex.

The: *Timeline* to achieve the targets is divided into the phases of the DMAIC. Make sure that the deadlines of the phases fit to the target deadlines.

At least you can specify the: *Team*. The list-boxes allow you and your Sponsor to build the team according to your specific needs. I entered of course the Sponsor and our process owner Chef du cookie as an accountable. The Sponsor and the accountable of the project should be managers with the power to make decisions about resources and the implementation of measures. It is a good idea to also invite a person from (financial) controlling to verify the monetary impact of the implemented measures later in the CONTROL phase. Additionally, I entered an important supplier and two selected external customers in this field too. Next you can list the members of the project core team. You can enter your name as Green or Black Belt and invite subject matter experts of the process and output for your support.

Depending on the cultural habits of your company, you, your Sponsor, as well as the controller could sign this charter like a contract. That can increase the binding nature of this agreement.

Tasks

Business-Projects: Please prepare the Project-Charter for your project.

Standard-Project: For your orientation some fields of the project scope are already specified. You can customize these entries according to your needs. And you should complete the remaining entries of the Project-Charter. Select the three most important problems and define the corresponding smart targets. I also hope that you will have a responsible manager of the garbage disposal service/city hall as a *Sponsor*. Maybe an *expert* of the operative service can assist you. Both are not obligatory, but it could make your work easier and more professional. You might also get support from established cleanup organizations of your country. Just browse the *World Waste Platform* (https://opendata.letsdoitworld.org/#/download) or search for information at the *World Cleanup Day* (https://www.worldcleanupday.org/).

6.4 Stakeholder Communication[4]

The Stakeholder Communication does also not belong to the problem-solving tools of the DMAIC. It is a tool for organizing the project and can help you to win and increase support by the people who are directly or indirectly affected by the targets of the project and their consequences. First, I will briefly explain the relevance of the topic and then introduce the tool.

Attitudes depend on many influences, e.g., experiences, beliefs, and values. Like the attitudes in other areas, one can assume that the range of possible attitudes of the affected people for or against your project is normally distributed. This means that some people will support your project, others will resist your project and most of them will be neutral. One could exaggerate the concrete behavior that follows from these attitudes into: *Enthusiasts*, *companions*, *spectators*, *underground fighters*, and *open enemies*.

Perhaps your project's stakeholders are exclusively *enthusiasts*. Or maybe you feel resistance in your environment against the project. Then you can and should do something about it. Let us take a closer look. Where does the resistance come from? And what can you do about it?

First, we can assume that the stakeholders of your project do not know that it is necessary and meaningful. They do not see the need for improvement; they only know Six Sigma superficially and have had bad experiences with change. People are

[4]Video 19 (https://www.youtube.com/channel/UCShsy50l5dFsoSXU__hBKSA).

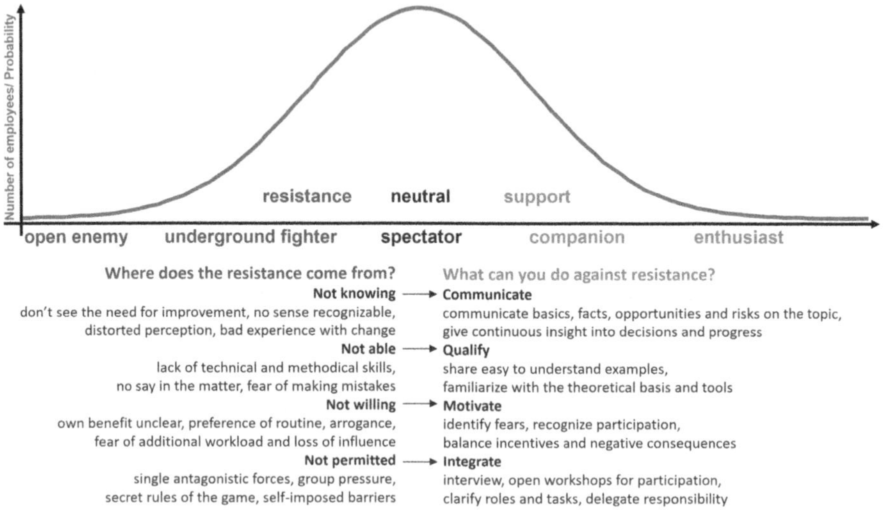

Fig. 6.10 Stakeholder Communication: Assume that the resistance to and support for your project is normally distributed. Take any expression of resistance seriously—it contains the key to gaining support

fundamentally skeptical about things they do not know. Therefore, you could communicate the basics of Six Sigma, give facts, address opportunities and risks on the topic, and give continuous insight into decisions and progress of the project (Fig. 6.10).

If we have knowledge of a subject, then on the next level we may fear that we lack the ability to cope with its possible technical and methodological requirements. We do not understand the Six Sigma language, cannot speak competently, and are afraid to make mistakes and embarrass ourselves. You could also qualify the stakeholders without calling it that or being a know-it-all. Repeatedly teach easy-to-understand examples, as well as the theoretical basics and tools, always in small doses.

As idealists, we are committed to topics that make sense to us. If we see no sense or benefit for ourselves, we remain neutral. Many people prefer familiar routines, face their own deficits with arrogance, fear additional workload and the loss of influence. This is about motivation. And we need to be able to answer the implicit question of everyone involved: "What's in it for me?" So, investigate what incentives and negative consequences could exist for the skeptics and invite them to participate directly or indirectly in the project to help shape it.

Antagonistic forces can prevent participation in the project or there can simply be doubts due to group pressure, secret rules of the game, or self-imposed barriers to participate. Of course, you cannot and should not force anyone to participate. But you can interview them, open your workshops for participation, clarify roles and tasks, and delegate responsibility.

You do not have to get every stakeholder excited about your project. It is probably enough to move the normal distribution curve just a little bit to the right.

Listen carefully to all statements about resistance, find the symptoms of resistance, and work on the causes. Avoid the pressure you feel to answer with back pressure. This can lead to an escalating dispute. Stay friendly and enjoy every little concession. Give the skeptics time and find the next opportunity for the next little concession.

For the tool first, make a list of your project's stakeholders. These are all people who are affected by your project and can influence its success positively or negatively. Some of these people are already listed in your Project-Charter. Other important stakeholders can also be outside the team, e.g., your boss, if he or she is not listed in the team.

Now focus on one of these stakeholders. I choose Mrs. Sponsor and selected the target: Y_01—Improve taste of cookies (grade of 1, 5 in customer rating). Now I assign Mrs. Sponsor the pseudonym *Spo*, which will ensure the anonymity of this person in the related Stakeholder chart.

Next, I rate her interest in target achievement. Our Sponsor is convinced of the project and is maximally interested in achieving its targets. As the responsible manager, she also has a very great influence on the achievement of the targets. The tool provides seven fields with which you can define the medium and the frequency of communication. Types of communication include: Newsletter, eMail, personal talk, team discussion, DMAIC phase steering, and the final presentation (Fig. 6.11).

I plan to regularly create a *newsletter* on the project's progress. And as frequency I choose: *monthly*. Furthermore, I would like to inform the Sponsor *weekly* by *eMail* about the project and hope for a *personal discussion*, at least *every 2 weeks*.

Stakeholder-Analysis and Communication-Plan

			Interested in …	Power to support the …	How do you want to win the support of this Person?					Degree of your Influence on this Person	Rank (Power x Interest x Influence)
Select one of your targets	Who in the company is positively/ negatively affected by the achievement of this target? (Name)	Pseudonym	… target-achievement		Type of communication	Frequency	Type of communication	Frequency	Type of communication		
Y_01 | Improve taste of Cookies (grade of 1,5 in customer rating)	Mrs Sponsor	Spo	10	10	newsletter	monthly	eMail	weekly	personal talk	5	1
Y_05 | Reduce consumption of ingredients and energy (- 20%)	Chef du Cookie	CdC	1	9	newsletter	monthly	eMail	weekly	personal talk	7	2
Y_03 | Adapt cookies to tin diameter (99% yield)	Mr D.E. Fender	DEF	10	1	newsletter	monthly	eMail	weekly	personal talk	1	4
Y_01 | Improve taste of Cookies (grade of 1,5 in customer rating)	Mr A. Pathetic	AP	2	2	newsletter	monthly	eMail	weekly	personal talk	3	3

Types of communication to win and increase support for your project:

- Newsletter,
- eMail,
- Personal talk,
- Team discussion,
- DMAIC phase-steering and the
- Final presentation

Fig. 6.11 Stakeholder Communication: Continuous communication about and integration into the project supports its implementation. The communication plan can help to win the support of important stakeholders

Furthermore, I invite Mrs. Sponsor to the *team discussion every 2 weeks*, as well as to *all workshop dates*. It is not necessary for her to join every meeting, but the invitation encourages her attitude and ensures transparency. It would be nice if she could at least come to the first workshop to explain the importance of the project to the team members. The *steering dates* are obligatory at the end of each DMAIC phase, in which the Sponsor decides on the successful completion of a phase or on necessary corrections. And of course, I also invite the Sponsor to the *final presentation* of the project.

Now I repeat this procedure for all other stakeholders and, depending on *interest* and *influence*, I determine the type and frequency of communication individually. For example, our Chef du cookie as an accountable has the power but little interest in changing her baking style, because she likes to nibble on the ingredients extensively.

Finally, I estimate the extent of my personal influence on these stakeholders. As a result, the tool provides a ranking of the stakeholders based on the multiplication of: *Power* × *Interest* × *My Influence* (Fig. 6.12).

This result is more differentiated in the bubble chart: Stakeholder. Here the stakeholders are arranged by their pseudonyms on the dimensions of: *Interest* and *Power*. And the size of the bubbles reflects: *My estimated influence* on them. The chart is divided into four quadrants: *Apathetics* with low interest and low power, *Defenders* with high interest but low power, *Latents* with low interest and high power, as well as *Promoters* with high interest and high power.

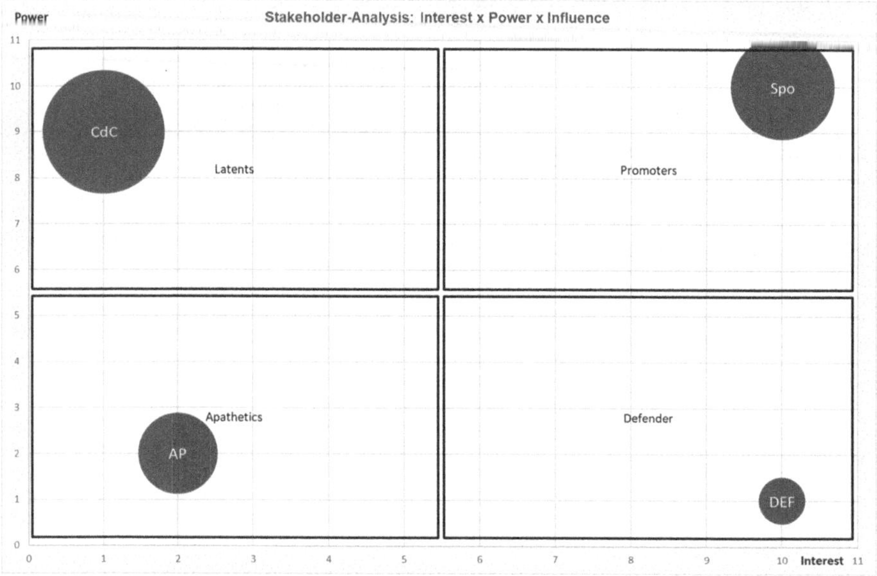

Fig. 6.12 Chart-Stakeholder: Stakeholder represented according to dimensions: Interest, power, strength of influence on the stakeholders (size of bubbles), and characterized by the four types: *Latents*, *Promoters*, *Apathetics*, and *Defender*

Your most important task could now be to positively influence the *Latents* for your project. Therefore, review whether your communication plan is appropriate especially for these important stakeholders.

Finally, please make the appropriate entries in your calendar. And if you like, you can already design the *newsletter*.

If you now have completed all tools of this phase, please prepare your Project-Story-Book for the DEFINE steering, first with your Master Black Belt and then with your Sponsor. As a Green Belt Certification candidate please check the details in the Certification-Path.

Questions (SC = Single Choice; MC = Multiple Choice)

Video 19/ Ch. 6.4/ #1 (SC)	Please assign the appropriate activities to the possible causes of resistance to your project.				
		Communicate	Qualify	Motivate	Integrate
1	Fear of making mistakes				
2	The need for improvements is not seen				
3	Actual/presumed group pressure against changes				
4	Additional workload/ less influence expected				
4	points				

Video 19/ Ch. 6.4/ #2 (SC)	Please assign the stakeholders to the different types according to their interest and power to support your project.				
		Defender	Apathetics	Latents	Promotors
1	high interest and high power				
2	low interest and high power				
3	high interest and low power				
4	low interest and low power				
4	points				

Tasks

Business-Projects: Please prepare the Stakeholder-Communication for your project. Copy the results of all DEFINE tools in your Project-Story-Book and comment on the results. Share the newsletter about your project also with your Master Black Belt.

Standard-Project: Please prepare the Stakeholder-Communication for your project, if you were able to convince employees (Sponsor, expert) of the garbage disposal service. Copy the results of all DEFINE tools in your Project-Story-Book and comment on the results. Share the newsletter about your project also with your Master Black Belt.

MEASURE 7

7.1 Overview: Steps, Tools, Their Purpose, and Results[1]

In the DEFINE phase we identified and prioritized the problems (Y) of the outputs with the help of experts. In the MEASURE phase we will identify the triggering influences (x) of the problems in the input and in the process also with the help of experts. We will relate influences (x) to problems (Y) and formulate hypothesis about these relationships and differences. Additionally, we will operationalize the influences and problems for their measurement and collect the respective data for x and Y (Fig. 7.1).

We start with the Input-Analysis and focus on the inputs already defined in the SIPOC. We will specify our requirements on the inputs and classify them to one of the categories of quality, availability, and consumption. We will also specify their deviations from the requirements and estimate the probability of their occurrence.

We continue with the Process-Mapping-Analysis and divide the process steps of the SIPOC into activities—from the first input (xi) to the last output (*Y*) of the SIPOC. We assign intermediate inputs and outputs to the activities as well as their associated methods (xm) and resources (xr). This increased transparency allows us to identify the negative triggering influences of the methods (xm) and resources (xr) in the activities on the problems of the outputs. Finally, we will estimate the probability of the occurrence of negative influences (xm; xr).

In the C&E-Matrix the experts estimate the impact strength of the relationships of the influences (xi; xm; xr) on the problems (*Y*). Then we will identify the strongest influences on the problems, as basis for statistical hypothesis. Finally, we will check the overall determination of each problem by the influences, to evaluate whether really all problems can be traced back to their triggering influences.

The Data-Collection-Plan prepares the measurement or collection of data of influences (*x*) and problems (*Y*). At first, we operationalize the influences (xi; xm;

[1]Video 20 (https://www.youtube.com/channel/UCShsy50l5dFsoSXU__hBKSA).

© Springer Nature Switzerland AG 2019

R. Hutwelker, *Six Sigma Green Belt Certification Project*, Management for Professionals, https://doi.org/10.1007/978-3-030-31915-1_7

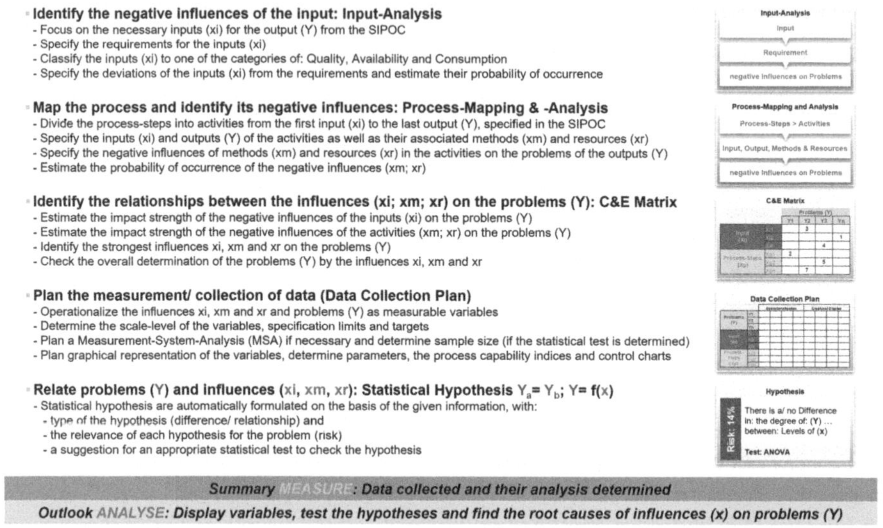

Fig. 7.1 Overview MEASURE: From the identification of influences on the problem to related hypotheses

xr) and the problems (*Y*) as measurable variables. Then we determine the scale level of these variables, we define specification limits and targets, and plan a Measurement-System-Analysis (MSA), if necessary. We determine the sample size and plan the graphical representation of the variables, determine parameters, the process capability indices, and suitable Control-Charts.

Finally, we transfer the relationships between the problems (*Y*) and their influences (xi; xm; xr) into the framework of statistical Hypotheses. sigmaGuide automatically formulates these statistical hypotheses based on the given information. sigmaGuide also calculates the relevance of each hypothesis for the problem as *Risks* and suggests an appropriate statistical test to check the hypothesis.

The purpose of the tools of MEASURE phase is thus to prepare the data collection, and to determine the methods for their analysis. These methods are then applied in the ANALYSE phase, in which we finally want to identify the root causes (x') of the triggering influences (x) of the problems (Y).

7.2 Input-Analysis[2]

In the Input-Analysis the negative influences of the inputs (xi) on the defined problems of the outputs (Y) are identified. In the Process-Mapping-Analysis, we will do the same for the negative influences of the activities (xm, xr). Please remember, that a project field, in which only negative the influences of the inputs

[2]Video 21 (https://www.youtube.com/channel/UCShsy50l5dFsoSXU__hBKSA).

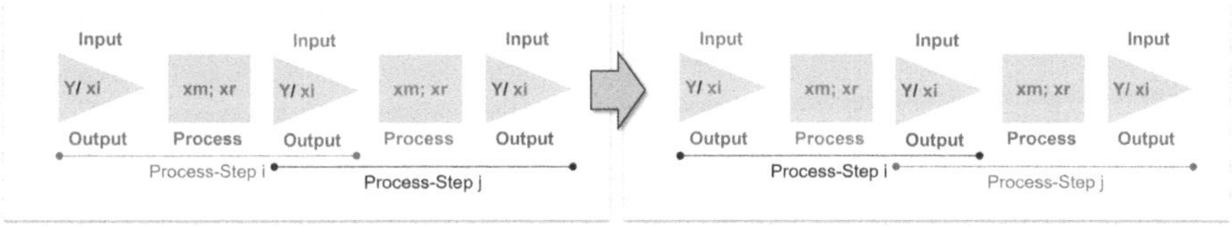

Fig. 7.2 Selection of the appropriate process field: If only negative influences from the input (xi) trigger the defined problems (Y), then the process field of the SIPOC must be shifted upstream, so that these influences (xi) become the problems (Y) of the upstream process step

trigger the problems of the output, might be incorrectly tailored, because typically you cannot improve the inputs. Especially if we cannot change their external impact on the process steps of our SIPOC. The SIPOC would have to be shifted upstream so that our critical inputs become the outputs of the upstream process steps (Fig. 7.2).

In the Input-Analysis we focus on the external inputs of SIPOC, specify our requirements and deviations from the requirements, and evaluate the frequency how often these deviations occur. In SIPOC, the process *bake cookies* has already been divided into its core process steps. The inputs of these input–process–output sequences can come from different sources: inputs from suppliers that flow into the process (a) *from outside* and inputs that result in the process flow as (b) *outputs from a previous process step.*

For this Input-Analysis, only the external inputs are relevant. Internal inputs are treated as outputs and have already been included in the tool: Voice-to-Criticals. Please note that such internal inputs are presented in [brackets] in the list-boxes. External inputs are displayed in the list-boxes without brackets. And they are to be analyzed here.

Let us start with the first external input xi1: *Request (cookie-type and delivery-date).* What do we need from the request? Of course, a request should be "unambiguous, complete and understandable."

The next question asks for the category to which this requirement belongs? I choose *Quality* from the list-box, because it affects the quality of our cookies.

Which deviation from the requirement is problematic for the process? It is problematic if the request is "ambiguous, incomplete or not understandable." Do you remember the letter with the order in the video: "Cookies should taste like those from Grandma" (Sect. 2.3)?

How often does this negative influence "Request ambiguous, incomplete or not understandable" occur? Let us say in 5% of all requests (Fig. 7.3).

I have already entered three more inputs and their influences. The second input is the *Start-Signal*. The process requires the *Start-Signal* to be given "on time." This requirement belongs to the category of: *Availability*. A problematic deviation from the requirement is when the *Start-Signal* is given "too late or too early." And, as our baking experts say, this happens in 10% of the cases.

The third external input is the *Tin*. Here we have already a clearly specified requirement with: "Diameter target of 110 mm, its lower specification limit (LSL) is: 107 and its upper specification limit (USL) is: 113 mm." This requirement belongs to

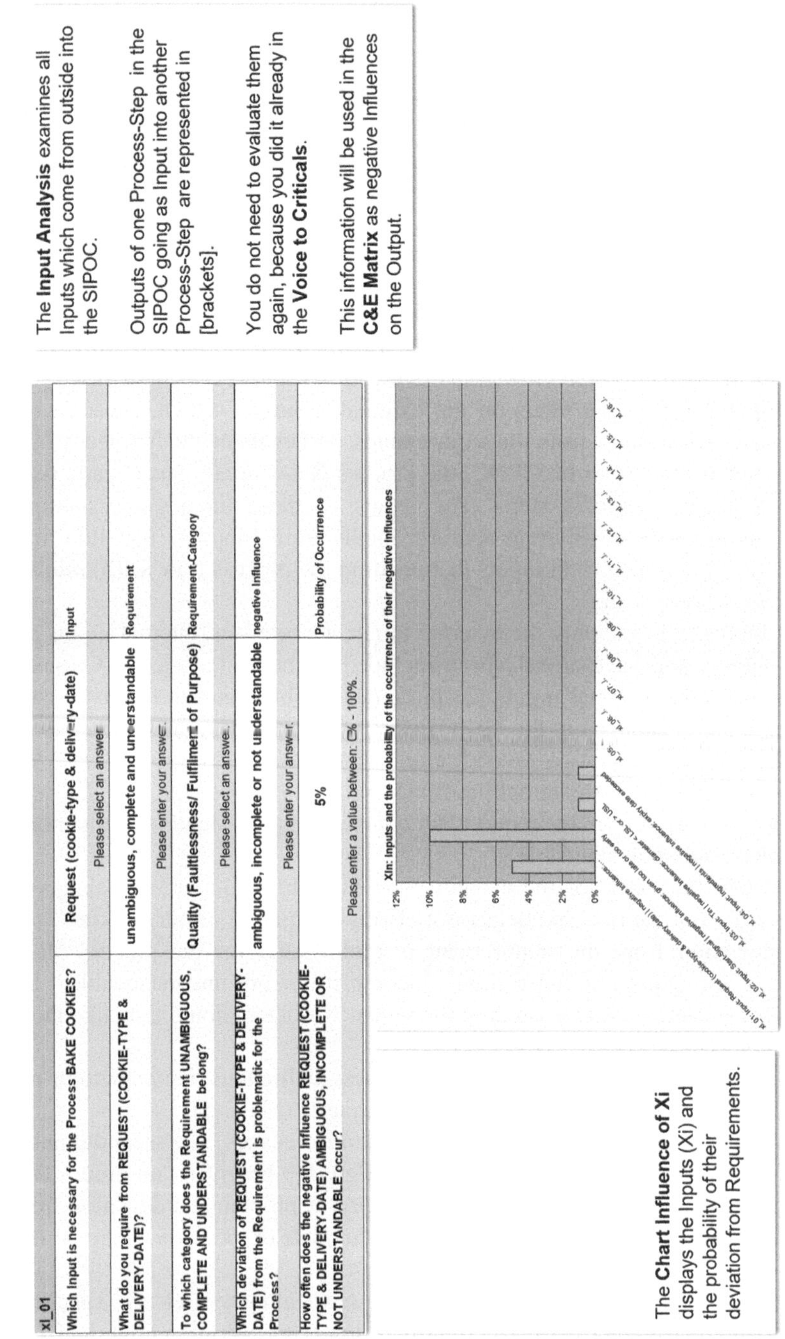

Fig. 7.3 Input-Analysis: Please answer the questions and get an overview of the negative influences of the inputs (xi) on the problems of the output (*Y*). They are essential for the C&E Matrix

the category of *Quality* again. A critical deviation from the requirement is given if the "diameter is smaller LSL or larger USL." This deviation happens in 1% of the cases.

The fourth inputs are the *Ingredients*. Here we demand that the quality attribute of the *expiry date must not be exceeded*. In fact, experts estimate that this occurs in 1% of cases.

Now we can switch to the *Chart-Influence of xi* to get an overview of all negative influences. The diagram shows all inputs entered and their probability of occurrence. As you can see, the xi2, the *Start-Signal* and its negative effect "being too late or too early," has the highest probability of occurrence.

We will now proceed to the Process-Mapping-Analysis and identify the negative influences of the activities on the attributes of our outputs. Please remember: Negative influences always affect attributes of outputs.

Questions (SC = Single Choice; MC = Multiple Choice)

Video 21/ Ch. 7.2/ #1 (MC)	What should you do if you could foresee that in your project only influences of the inputs trigger the defined problems?	
1	That's normal, no need to change anything	
2	I invite the supplier to join the project	
3	I shift the project field upstream so that the inputs become the outputs	
4	I inform my sponsor	
4	points	

Tasks

Business-Projects: Please prepare the Input-Analysis for your project by specifying the inputs of your SIPOC. This information is related to the C&E Matrix and to the Data-Collection-Plan and thus determines your measurements.

Standard-Project: Please prepare the Input-Analysis for your project by specifying negative influences of the inputs. This information is related to the C&E Matrix and to the Data-Collection-Plan and thus determines your measurements. For your orientation some entries of the tool Project-Definition are already prefilled. You can customize these entries according to your needs.

7.3 Process-Mapping/-Analysis[3]

The Process-Mapping/-Analysis details and supplements the SIPOC. We have defined the limits of the SIPOC in such a way that its core process steps from the chronologically last problem, upstream to the first relevant input, enclose all triggering influences (x) of all defined problems (Y). In the Process-Mapping part

[3]Video 22 (https://www.youtube.com/channel/UCShsy50l5dFsoSXU__hBKSA).

we will first detail the process steps into underlying discrete activities. In the Process-Analysis part we will then identify the triggers of the problems (Y) in the methods (xm) and resources (xr), as we already did it with the inputs (xi) in the Input-Analysis.

7.3.1 Process-Mapping

If you do not perform all process steps of the SIPOC yourself, between first input and last output, you should conduct a workshop with the employees who perform the activities of the process steps and know the negative influences from their daily work. If you first map and analyze the process on flip charts, then all participants can easily follow and contribute. Only after the workshop you should transfer the most important information of the flip charts into sigmaGuide.

I have seen many Process-Mappings of varying complexity. In a rather complex purchasing process, from ordering to delivery, we wallpapered all walls of the team room with flip charts of the Process-Mapping. In another, quite easy process, we only needed half of a flip chart for a tool change on a machine (Fig. 7.4).

In the workshop, I first note the name of the process in the first line of the first flip chart. Then I write the names of all departments involved in the process steps on stickers. Next, we order these departments in the first column in the chronology of their appearance in the process. This is the prerequisite for arranging the activities on the flip chart in the typical cascade from the upper left to the lower right side. Now we know the answer to the first key question:

1. *Who is involved in the process?* (Fig. 7.5)

Next, we will find out: *What is done?* To prepare the place for the activities, I draw horizontal boundary lines in the flip chart for each department. This type of process-mapping is called: *Swim-Lane-Diagram*, in contrast to the *Flow-Chart*, which is often used for official process documentation. But we do not want to document the process; we would not even consider decisions that are usually marked as diamonds in a flow chart.

In our bakery, the customer is on top of the department list and I ask:

(2a) *What is the customer typically doing first?*

We already know the answer from SIPOC: "transmit request for cookies." I note this activity on a sticker and post it to the right of the first department. Make sure that you always describe activities at least with a verb and a noun. Because only verbs can express activities and the noun represents the object of the certain activity. To really understand the process, it is important to describe the activities as concisely and precisely as possible. This is not always easy for me and I sometimes need several attempts to really describe what happens. But this is a good exercise for the: Learning to see. Thus, please try it.

Key questions for the workshop

Involved departments

1. Who is involved in the process?

Activities

2.a What is typically done first?

2.b What happens next? (variant)

Input & Output

3. Which input is necessary for the activity?

4. Which output results from the activity?

Methods & Resources

5. Which resources execute or support the execution the activity?

6. Which methods guide the execution?

Influences & Frequency

7. Which influences from the methods and resources on or by the activities negatively affected the quality, the availability of the output or the consumption and waste of input and resources in the past?

8. How often (%) did the most relevant and most important of these influences affect the output in the past?

Stay to these question, repeat them, do not vary them, because you might get a different answer than intended

Fig. 7.4 Process-Mapping-Analysis: Key questions and procedure for the workshop with the subject matter experts of the process. Please apply this standard and make your own experiences before changing it

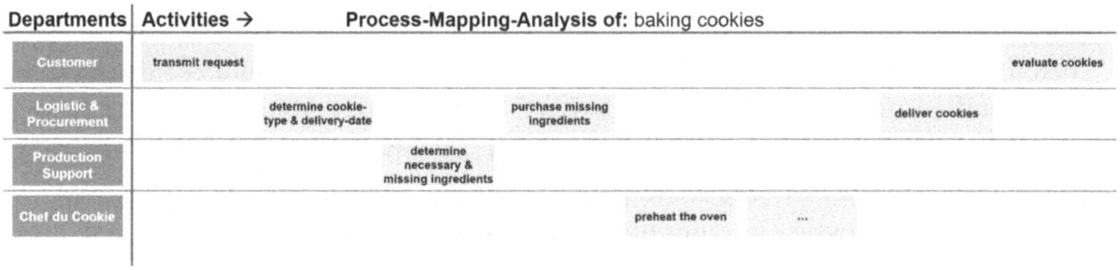

Fig. 7.5 Process-Mapping: Draw swim lanes for the involved departments and specify their activities with stickers on flip charts, one after the other, from left to right, without overlapping, and more detailed than in the previous SIPOC

To continue the Process-Mapping I ask a variant of the second question:

(2b) *What happens next?*

And I also write the answer: "determine cookie-type and delivery-date on a sticker." This is the first activity of the department: *Logistic and Procurement.* And I put the sticker of this second activity in the corresponding swim lane, right of the first activity and below the first swim lane. This sequential order of activities from left to right illustrates the chronology of the process. And it ensures that we later can examine *activity* by *activity* individually for its negative *influences.*

What happens next? The department: *Production Support* "determines necessary and missing ingredients." The following activity "purchase missing ingredients" will again be carried out by *Logistic and Procurement.*

We do not need a second swim lane for the second activity of this department but stick the note in its existing swim lane. The repeated involvement of the same departments, visible by the interruption of the sticker cascade, reveals potential in the interfaces and in the sequence of activities, later in the IMPROVE phase.

Next, our Chef du cookie: "preheats the oven" and continues the chronology of activities in the cascade of notes. I continue to collect the activities up to "deliver cookies," which directly leads to the last problematic output: Cookie *(delivered).*

7.3.2 Process-Analysis

Now I prepare the Process-Analysis part for the collection of further information for the analysis of the activities. In the next swim lane, the *inputs* of the activities can be specified. This is not strictly necessary for the analysis but can support the orientation in the diagram. Thus my third question is:

3. *Which input is necessary for the activity?*

Strictly necessary for the analysis are, however, the *methods* and *resources* used for the specific implementation of the *activity.* Any negative *influence* on and from a specific *activity* subtracts something from an assumed best practice *activity* to generate a best practice output in the targeted quality, availability, and consumption.

It is again not necessary to enter the *outputs* of the activities in the next swim lane, although this, as with the input, can again facilitate orientation in the diagram. My fourth question is accordingly:

4. *Which output results from the activity?* (Fig. 7.6)

In the next three lines, the negative *influences* of the *methods* and *resources* on or from the *activity* must be specified. And in the last row, the *frequency* must be estimated, how often the negative *influence* affects the activity.

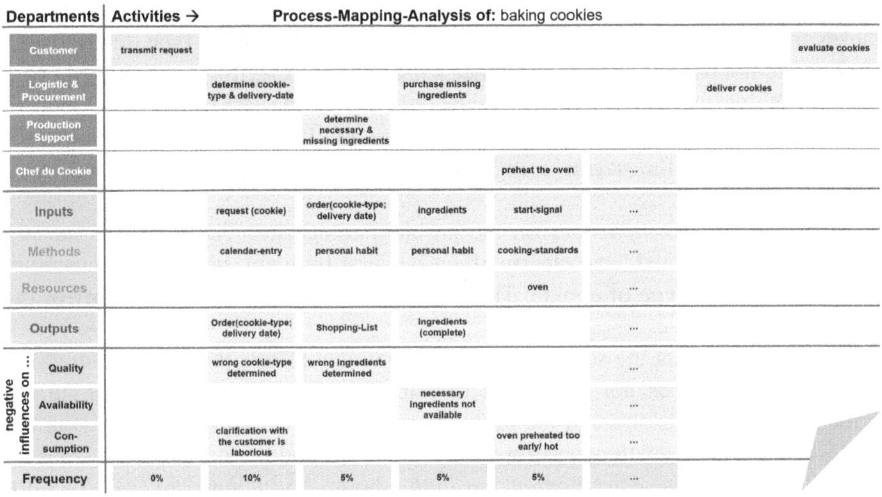

Fig. 7.6 Process-Analysis: Assign inputs, outputs, methods, resources to the activities, identify the triggering influences in the methods (xm) and resources (xr) of the problems (*Y*) and estimate their frequencies

Let us start the analysis. The first activity: "transmit request for cookies," is executed by the *Customer*, and we do not know and do not care about the *influences* in activities of customers.

After receiving the input: *Request (cookie)*, already evaluated in the Input-Analysis, our bakery: "determines the cookie-type and delivery-date." Now I ask the participants two questions about the *methods* and *resources*:

5. *Which resources execute or support the execution of the activity?* and
6. *Which methods guide the execution?*

The *method* of the focused activity in our video is a simple "calendar-entry" (Sect. 2.3). The *resources* are: Laura—in her role as *Logistic and Procurement Service*—a "calendar" and a "pencil." I only enter these trivially given *resources* if they are a source of a negative *influence*.

The next question is about the negative *influences* and I ask my participants:

7. *Which influences from the methods and resources on or by the activities negatively affected the quality and the availability of the output or the consumption and waste of input and resources in the past?*

And I get and note the answer that very seldom the "wrong cookie-type" was determined and that sometimes in the past "laborious clarifications" with the customer were necessary.

Now I check whether these *influences* are direct triggers of problems and not root causes from the end of the causal chain. I accept "wrong cookie-type determined," but I would not have accepted "customer's handwriting illegible," which could be a cause of the wrong determination. Please only accept the last trigger in a chain of causes that directly *influences* the activity negatively. It must be always obvious for you that and how the negative *influence* can affect the activity and leads to a problem.

Now I assign the *influences* on the problem categories. The *influence* "wrong cookie-type determined" can clearly affect our first problem, because the customer expects a different type of cookie. Therefore, it is an *influence* on the quality of the output.

It is noticeable that the second *influence* negatively affects the time consumption of resources but does not affect any of our defined problems. However, in the workshop the experts stressed the high amount of effort involved in clarifying the orders in the past. Now the question arises whether the additional problem: Cookies *(boxed) working time consumption too high*, should be included in the Voice-to-Criticals tool, to be able to continue its analysis. It is quite possible that you also find negative *influences* in your activities that have no relation to one of your defined problems. And it would be perfectly normal to adjust the Project-Charter retroactively, together with the Sponsor, to add or change entries. I do not do that here, but ask the last question for this activity:

8. *How often (%) did the most relevant and most important of these influences affect the output in the past?*

And my participants estimate the frequency of clarifying a request at 10% of the cases.

We switch to the third activity: "determine necessary and missing ingredients," which transforms the *Order (cookie-type; delivery date)* into the *Shopping-List.*

My question about the *methods* that control this activity leads first to amazement and then to the statement that it is very simple and that every employee follows his or her personal habits. As a method I therefore write: "personal habit." And of course, you know that personal habits can lead to variability in the *activity* and thus to variability in output.

My question about the concrete negative *influences* is answered by the participants by the fact that in the past sometimes the wrong ingredients were determined. And they estimate its frequency at 5%. This is not so often, but the negative *influence* of "wrong ingredients" on the taste of the cookies and thus on our first problem is obvious. In the C&E matrix we will later evaluate the strength of this relationship.

In the fourth activity: "purchase missing ingredients," the negative *influence*: "necessary ingredients not available" occurred in about 5% of the cases. And there is probably a relationship to the second problem: Cookies *delivered too early/late*.

In the fifth *activity*, the oven is: "preheated too early and/or too hot" in 5% of the cases. This affects the energy consumption and thus our third problem.

So, I continue for all remaining *activities* and collect the negative *influences* and their frequencies up to the "delivery of the cookies" and their "evaluation by the customer."

Finally, I reviewed the prepared Process-Mapping-Analysis with the experts, corrected mistakes and rearranged or added missing information, which is very easy with the stickers.

7.3.3 Procedure of the Workshop

Now think about your own workshop. And what do you think of these suggestions? (Fig. 7.7)

1. First, keep calm. A certain amount of inner tension is good, because it indicates that your necessary mental energy is mobilized. However, too much nervousness

Procedure for the workshop

1. Keep calm – if you get excited: speak slower, breathe evenly, deeply
2. Invite participants and inform the sponsor
3. Ask for a guided tour of the process, get the workshop materials, prepare the room
4. Take care of your participants
5. Take the guided tour of the process
6. Repeat the following microprocess with for each of the key questions (left) for each activity, until you have received a valid answer:
   a) ask one of the standard key questions,
   b) check the answer,
   c) summarize relevant aspects of the answer and
   d) let the participants check your understanding of the answer
7. Note the relevant information of the answers on different stickers each:
   a) **Activities:** describe them as precisely as possible with: verb & noun
   b) **Input, Output, Methods & Resources**: write their names or "personal habits", if e.g. methods are missing
   c) **negative influences:** Identify the last direct influence, i.e. the last trigger in the causal chain that causes or contributes to a defined problem, check: **If influence occurs, then problem arises/increases?**
8. In between thank your experts for their participation and their answers
9. Take a picture of yourself and the team in front of your workshop flip charts

Fig. 7.7 Process-Mapping-Analysis: Procedure for the workshop with the subject matter experts of the process. Please apply this standard and make your own experiences before changing it

can also be an obstacle. If I get too excited, I try to speak slower, breathe more evenly and deeply. And this helps me to calm down.

2. Invite participants and inform the Sponsor.
3. Ask for a guided tour of the process, get the workshop materials, and prepare the room.
4. Take care of your participants—the better they feel, the better your results.
5. Take the tour of the process, real on the shop floor or virtually through the screen of a software, no matter if you know the process or not. Let the experts show you inputs, outputs, methods, resources, activities, and negative influences. And if allowed, take photos and screenshots as a reference for the workshop.
6. In the workshop repeat the following microprocess until you have received a valid answer to the eight questions for each activity:
   (a) Ask one of these standard questions and do not vary them, because you might get a different, invalid answer.
   (b) Check the answer.
   (c) Summarize relevant aspects.
   (d) Let the participants check your understanding of the answer.

   You do not need to know everything about the process. But you need to understand the workflow, know its influences and know how they are impacting. Concentrate on the questions and so lead the workshop again with the key questions if you get off the subject.
7. Note the relevant information on a sticker:
   (a) For the Process-Mapping: note the activities in the format: verb and noun. Find the verb and noun that most accurately describe the activity. This ensures that you describe the actual as-is process. Do not rely on given process descriptions and record wishes mentioned for the should-be process on an additional flip chart. In detailing the core process steps into individual activities, you need to find the balance between too specific and too general descriptions.
   (b) For input, output, methods, and resources: Simply write their names on different stickers each for the flip chart. If there is a method missing, then write: "personal habits."
   (c) For negative influences: Identify a direct influence, i.e., the last trigger in the causal chain that causes or contributes to the defined problems. Check the relationship between influence and problem. This relationship must be rationally comprehensible, such as the influence of: "too early preheating of the oven" on our problem of "too high energy consumption."
8. In between thank your experts for their answers. Nobody likes to talk about mistakes, but we are overall talking about mistakes in our project. It is important that it is never about guilt, that you do not look for anyone to blame and prevent the allocation of guilt, because we want to solve problems and create a good climate in our personal relationships.
9. Finally, take a picture of yourself and the team, in front of your flip chart, as a reminder for you and as a workshop confirmation for us.

7.3.4 Results

After the workshop you can copy the results of the flip charts to sigmaGuide. First select the executing departments from the upper left list-boxes with the suppliers and customers of the SIPOC. Then transfer the activities, inputs, outputs, methods, resources, and negative influences with their probability (Fig. 7.8).

The number of activities on your flip chart is probably higher than the number of core process steps of your SIPOC. sigmaGuide can, however, only take up a maximum of 20 activities in total. When you transfer activities from the flip chart, it may be necessary for you to combine, e.g., two activities in one column. Important here is, however, that all relevant negative influences are entered and assigned to their activities (Fig. 7.9).

If you transferred all information from the flip chart to the tool, you can go to the related chart. In the chart: Influence of xm and xr, you get an overview of the activities with their negative influences and their probability. As you can see, the eighth activity, where our Chef "tests the chocolate," has the highest probability to occur.

With this first workshop you will have completed a significant part of your project again and you can enter the results in your Project-Story-Book.

Questions (SC = Single Choice; MC = Multiple Choice)

Video 22/ Ch. 7.3/ #1 (SC)	Please bring the following steps to completing the tool: Voice-to-Criticals in their correct order.					
		1st step	2nd step	3rd step	4th step	5th step
1	What are the inputs and outputs of the activities?					
2	Which departments are involved?					
3	Which influences of methods and resources trigger the problems in which frequency?					
4	Which specific activities are carried out?					
5	What are the methods and resources of the activities?					
5	points					

Video 22/ Ch. 7.3/ #2 (MC)	What is the purpose of the Process-Mapping/ -Analysis?	
1	Creation of a certifiable document	
2	Identify the root causes of the problems	
3	Identify triggers of the problems	
4	Get an overview of the involved departments	
5	Identify loops, critical interfaces and unnecessary responsibility changes	
5	points	

Process-Mapping-Analysis of the Process: bake cookies

Who ...	... does what?	1. Activity	2. Activity	3. Activity	4. Activity	5. Activity	6. Activity	7. Activity
		Please specify the Process-Steps in detailed Activities the format: Verb + Noun (e.g.: weigh Ingredients)						
1. Process-Step	Customer (Child)	transmit request for cookies						
2. Process-Step	Logistic & Procurement Service		determine cookie-type & delivery-date		purchase missing ingredients			
3. Process-Step	Production Support			determine necessary & missing ingredients		preheat the oven		
4. Process-Step	Chef du Cookie						clean the workplace	weigh ingredients
Input:	Which Inputs are necessary to start the Activity?	...?	Request (cookie)	Order (cookie-type; delivery-date)	Ingredients	Start-Signal	...?	Ingredients (complete)
Methods:	Which Instructions/ Rules direct how to perform the Activity?		calendar-entry	personal habit	personal habit	cooking-standards	kitchen-standard	personal habit
Resources:	Which Equipment/ Machines/ Tools operate or support the Activity?					oven	cloth and detergents	kitchen-scale; measuring-cup
Output:	Which Output results from the Activity?	...?	Order (cookie-type; delivery-date)	Shopping-List	Ingredients (complete)	...?	Workspace (clean)	Ingredients (weighed)
Which Influences of the: - Methods and - Resources negatively affect:	... the Quality (Faultlessness/ Fulfilment of Purpose) of the Output?		wrong cookie-type determined	wrong ingredients determined				weight of ingredients too low
	... the Availability (right Quantity just in Time) of the Output?		delay in case of unavailability of the customer		necessary ingredients not available			
	... the Consumption and Waste of Input and/ or Resources?		clarification with the customer is laborious			oven preheated too early/ hot		
How often are the Activities affected by these negative Influences?		0%	10%	5%	5%	5%	0%	30%

Fig. 7.8 Transfer the information of the Process-Mapping-Analysis after the workshop to sigmaGuide. Most important are the negative influences of the methods (xm) and resources (xr) and their probability on the defined problems. They are essential for the C&E Matrix

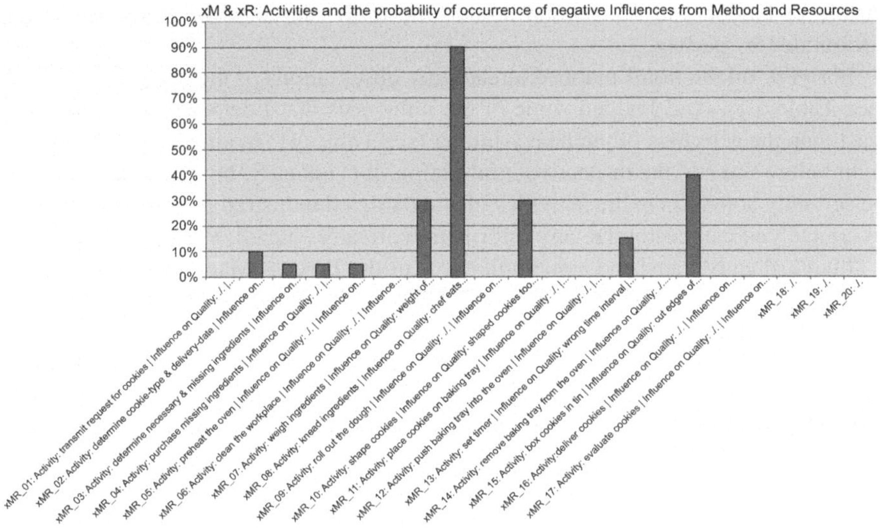

Fig. 7.9 The Chart-Influence of x-process graphically displays the activities, their negative influences (xm, xr) with their frequency on the problems of the output (*Y*)

Tasks

Business-Projects: Please prepare the Process-Mapping for your project. Identify the methods that guide and the resources that execute the respective activities. Identify the negative influences on the activities and estimate their frequency. This information is related to the C&E Matrix and to the Data-Collection-Plan and thus determines your measurements.

Standard-Project: The SIPOC of this project contains only a few core process steps. And it might be difficult for you to resolve each step into detailed activities here, because the activities of littering depend on a sequence of individual decisions and habits.

The two littering types (occasional-litterer; heavy-litterer) presented in the SIPOC offer a classification system for observations, for the construction of the questionnaire and for the mapping and analysis of activities.

You might, e.g., observe these different behaviors: One consumes his fast food near a trash can and throws the packaging into it. Another looks around, searches for a trash can, goes there, and throws the packaging into it. Another one puts the garbage into his own bag and carries it away. Another one is in a hurry and drops his garbage immediately after consumption. Another one, perhaps in a group of people who have some fun, throws the garbage in a wide arc on the floor. In these cases, the triggering influence lies in the attitude of the people, which could be well described by the littering types. But the threshold of littering could also be increased or decreased by external influences that are the responsibility of the city: The nearest garbage can could be far away or near. It can be empty or overcrowded, clearly visible or hidden and the ground around the trash can could be clean or already surrounded by garbage.

For your orientation I mapped already my observations of the occasional-litterer (see Table 6.1). And I added some of the influences that I detected. You should at least map the activities of the heavy-litterer (see Table 6.1) in an adjacent swim lane right below and add the respective information. It is not easy, but if you do, you will be certainly able to map the activities of standardized industrial processes later. And of course you can customize any information already entered. Please try it out.

I have already translated some influences of the Process-Mapping/-Analysis into questions (Q2, Q3, Q4, Q5, Q6, Q7). The corresponding questionnaire (see Fig. 7.23), introduced in Sect. 7.5.2, can help you to collect the necessary data.

In the following C&E Matrix you can already evaluate the relationship between these influences—represented in the questions—and the problems. And later you can operationalize associated metrics in the Data-Collection-Plan.

7.4 Cause and Effect Matrix[4]

The C&E Matrix (Cause and Effect Matrix) links the identified influences of the Input- and Process-Mapping-Analysis (xi, xm, xr) as causes, with the defined problems (Y), their effects. Experts should now assess the strength of the relationships of each influence on each problem. The resulting influence–problem pairs (x–Y) serve as a basis for statistical hypotheses. Thus, the C&E Matrix represents the central function of Six Sigma: $Y = f(x)$ like no other tool.

For me, the C&E Matrix is the heart of DMAIC, because every relevant x–Y relationship will later be translated into a hypothesis, for which data is collected, statistically tested, and the significant relationships become the starting point for Root-Cause-Analyses. And you remember: The root causes are the interfaces for improvements.

sigmaGuide lists all defined problems in the columns of the matrix, together with the respective Kano category and the calculated severity (Voice-to-Criticals). In the rows, the matrix lists the influences of the inputs (xi) and activities of the process (xm, xr), together with their respective frequency and ranking (Input-Analysis; Process-Mapping-Analysis) (Fig. 7.10).

By default, all cells that relate influences and problems are empty. This means that the respective influence has no effect on the respective problem, corresponding to a 0% determination of the certain problem by the certain influence. In the other extreme, an influence strength of 100% means that the occurrence of an influence completely triggers the occurrence of the problem or at least always leads to increasing the problem. Thus, the strength of the *x–Y* relationship can vary between 0% and 100%.

[4]Video 23 (https://www.youtube.com/channel/UCShsy50l5dFsoSXU__hBKSA).

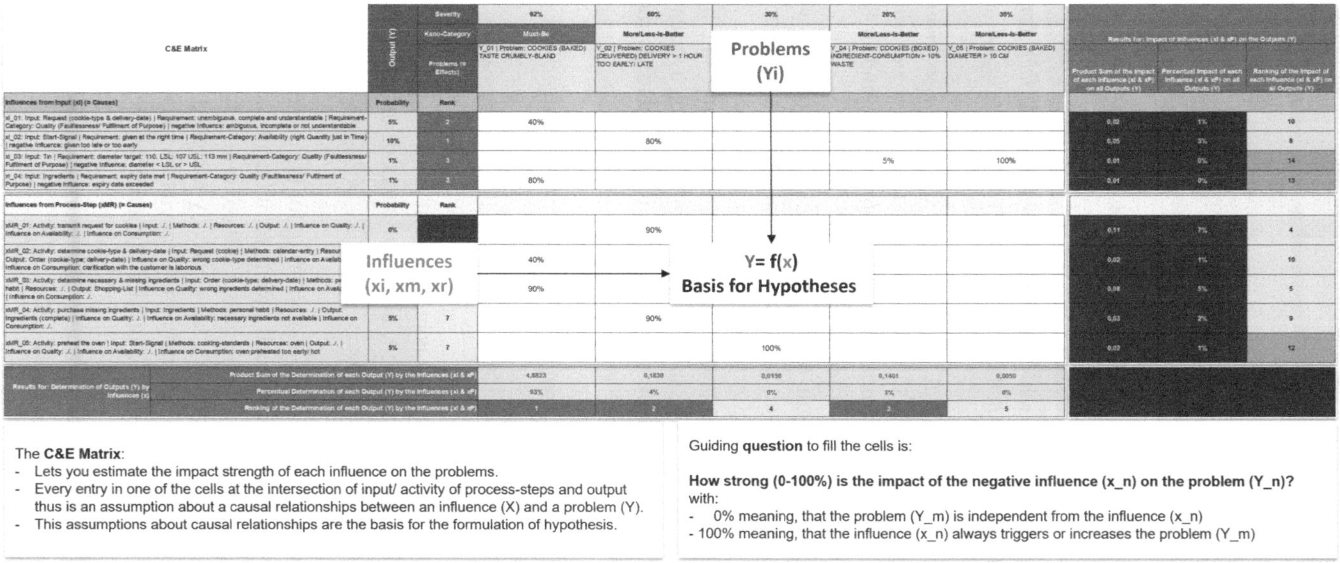

Fig. 7.10 C&E Matrix: Specify the impact strength of each negative influence in the inputs (xi) and activities (xm; xr) on the problems of the outputs (Y)

C&E Matrix	Output (Y)	Severity	92%
		Kano-Category	Must-Be
		Problems (= Effects)	Y_01 | Problem: COOKIES (BAKED) TASTE CRUMBLY-BLAND
Influences from Input (xI) (= Causes)	Probability	Rank	
xI_01: Input: Request (cookie-type & delivery-date) | Requirement: unambiguous, complete and understandable | Requirement-Category: Quality (Faultlessness/ Fulfilment of Purpose) | negative Influence: ambiguous, incomplete or not understandable	5%	2	40%
xI_02: Input: Start-Signal | Requirement: given at the right time | Requirement-Category: Availability (right Quantity just in Time) | negative Influence: given too late or too early	10%	1	

Fig. 7.11 Extract of the C&E Matrix with the relationship strengths between xi_01 and Y_01 (40%) and between: xi_02 and Y_01 (0%)

In our bakery the problem: Cookie *taste crumbly-bland (Y_01)*, for example, occurs completely independently of the time at which the *Start-Signal is given for baking (xi_02)* (Fig. 7.11).

In contrast, experts estimate that the problem: Cookie *taste crumbly-bland* (Y_01) occurs in 40% of cases if the input: *Request (cookie-type and delivery-date)* is *ambiguous, incomplete or not understandable* (xi_01) and in 80% of the cases, if the *expiry date of the ingredients is exceeded* (xi_04).

If you do not know the strength of the influences yourself, then you should involve the experts again. I have had good experiences with two approaches:

(a) Ask the experts to collectively assess each cell in a workshop, one by one, or
(b) Let the experts make the assessment independently of each other and then calculate the mean value for each cell from these individual assessments.

In any case, your question to the experts can be:

How strong (0–100%) is the impact of the negative influence (x_n) on the problem (Y_m)?

with:

- 0% meaning, that the problem (*Y_m*) is independent from the influence (*x_n*) and
- 100% meaning, that the influence (*x_n*) always triggers or increases the problem (*Y_m*).

The C&E Matrix is very large, and it can be tiring for the experts to evaluate each cell. To facilitate this work, an empty cell means an influence strength of 0%. So, you do not have to change anything in the matrix if there is no effect from an *x* to a *Y*.

Furthermore, in the previous tools we have already assigned the influences and problems according to their type of impact and the deviation to one of the categories quality, availability, or consumption. Logically comprehensible, quality problems are primarily caused by influences on quality and not by influences on availability or consumption. This specificity between trigger and problem can, therefore, simplify the overview and processing of the many cells.

The premature or delayed *start-signal for baking (xi_02)*, already categorized as an influence on availability, thus has more effect on the *delivery time of the cookies (Y_02)* than on their *taste (Y_01)* or their *energy consumption (Y_03)*.

Also make sure that you only rate *x*–*Y* pairs that are directly related and not the many possible indirect relationships between influences and problems. If, e.g., a delayed *start-signal (xi_02)* leads to the *expiry date being exceeded (xi_04)*, then this could have an indirect negative effect on the *taste of the cookies (Y_01)* and on the *disposal of inedible ingredients (Y_04)*. However, these indirect relationships should not be evaluated. Instead, the direct relationship of the *start-signal (xi_02)* to the *delivery of cookies (Y_02)* should be assessed, as well as the direct influence of the *expiry date (xi_04)* on the *taste of the cookies (Y_01)* and on the *disposal of inedible ingredients (Y_04)*.

Each influence can affect several problems and a specific problem can be triggered or increased by several influences.

After you have evaluated the *x*–*Y* relationships, sigmaGuide offers three results: (1) the overall effect of each influence on all problems, (2) the overall determination of each problem by all influences, and (3) risk of each influence–problem pair.

1. The last three columns of the matrix show the overall effect of each influence on all problems. In the third last column the product sum is calculated for each influence: Frequency of the influence *x* severity of the problem *x* the strength of the relationship between influence and problem. In the penultimate column, the product sum is transformed into the percentual impact of each influence on all problems. And the last column shows the impact ranking for all influences. Here you can distinguish between the highly relevant and the less relevant influences (Fig. 7.12).

2. The last three rows of the matrix show the overall determination of each problem by all influences. As with the influences, the product sum, the percentual determination, and the ranking are calculated for each problem. Here you can distinguish problems that are relatively well determined by influences from problems that are little or perhaps not at all determined by the listed influences. If so, you might go back to the Input- and Process-Mapping-Analysis to look for the specific trigger of this undetermined problem.

C&E Matrix	Output (Y)	Severity / Kano-Category / Problems (= Effects)	Results for: Impact of Influences (xi & xP) on the Outputs (Y)		
			Product Sum of the Impact of each Influence (xi & xP) on all Outputs (Y)	Percentual Impact of each Influence (xi & xP) on all Outputs (Y)	Ranking of the Impact of each Influence (xi & xP) on all Outputs (Y)
Influences from Input (xi) (= Causes)	Probability	Rank			
xi_01: Input: Request (cookie-type & delivery-date) | Requirement: unambiguous, complete and understandable | Requirement-Category: Quality (Faultlessness/ Fulfilment of Purpose) | negative Influence: ambiguous, incomplete or not understandable	5%	2	0,02	1%	10
xi_02: Input: Start-Signal | Requirement: given at the right time | Requirement-Category: Availability (right Quantity just in Time) | negative Influence: given too late or too early	10%	1	0,05	3%	6
xi_03: Input: Tin | Requirement: diameter target: 110, LSL: 107 USL: 113 mm | Requirement-Category: Quality (Faultlessness/ Fulfilment of Purpose) | negative Influence: diameter < LSL or > USL	1%	3	0,00	0%	13
xi_04: Input: Ingredients | Requirement: expiry date met | Requirement-Category: Quality (Faultlessness/ Fulfilment of Purpose) | negative Influence: expiry date exceeded	1%	3	0,01	0%	12

Fig. 7.12 Extract of the C&E Matrix with the overall impact of each influence (x) on all problems (Y)

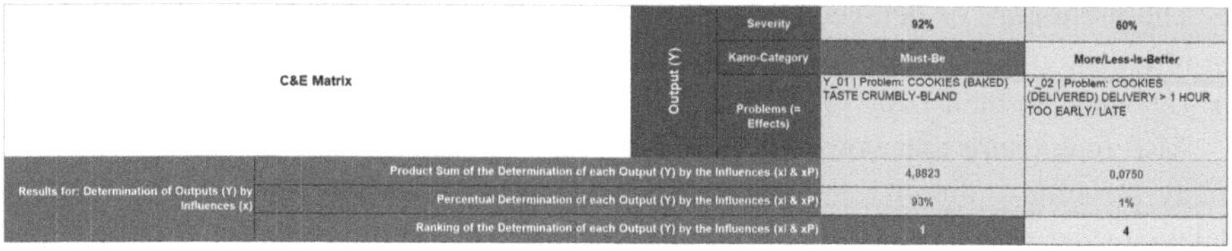

C&E Matrix	Output (Y)	Severity	92%	60%
		Kano-Category	Must-Be	More/Less-Is-Better
		Problems (= Effects)	Y_01 | Problem: COOKIES (BAKED) TASTE CRUMBLY-BLAND	Y_02 | Problem: COOKIES (DELIVERED) DELIVERY > 1 HOUR TOO EARLY/ LATE
Results for: Determination of Outputs (Y) by Influences (x)		Product Sum of the Determination of each Output (Y) by the Influences (xI & xP)	4,8823	0,0750
		Percentual Determination of each Output (Y) by the Influences (xI & xP)	93%	1%
		Ranking of the Determination of each Output (Y) by the Influences (xI & xP)	1	4

Fig. 7.13 Extract of the C&E Matrix with the overall determination of each problem (Y) by all influences (x)

3. The Risk of each influence–problem pair is the most important result. It is calculated from the available information and displayed in the: Chart Cause and Effect Heatmap. These risks express the basic function of Six Sigma: $Y = f(x)$ and represent the relevance of the resulting hypotheses (Fig. 7.13).

Each risk in each cell results from the multiplication of the frequency of an influence, the severity of its cause, or increased problem and the strength of the relationship between influence and problem. And these risks can vary between 0% and 100%. It is unusual to speak of risks in the context of Six Sigma. But we can consider each x–Y pair a risk, because risks are typically calculated based on the severity of an event and the probability of its occurrence.

If you zoom out of this table, you will get an overview of all risks in your process. Please note that the red fields indicate risks higher than 10%, the yellow fields indicate risks between 1% and 10%, and the risks of the green fields are below 1%. Thus, the colors serve to prioritize the corresponding hypothesis between x and Y, and you can later easily decide which hypothesis you want to further investigate.

But please carefully check the green fields, the underlying influences and their frequency, the problems and their severity, as well as the strength of their estimated relationship before you exclude them from further analysis (Figs. 7.14 and 7.15).

For a clear presentation, please transfer the important results of the C&E Heatmap into the summarized and simplified template of the C&E Matrix in the Project-Story-Book.

Questions (SC = Single Choice; MC = Multiple Choice)

Video 23/ Ch. 7.4/ #1 (SC)	In the C&E Matrix, influences (x) in the process are matched against problems in outputs (Y).			
		Relationship between x and Y	Overall influence of each x	Overall determination of each Y
1	What is evaluated in the cells of the C&E Matrix?			
2	What is the result per row?			
3	What is the result per column?			
3	points			

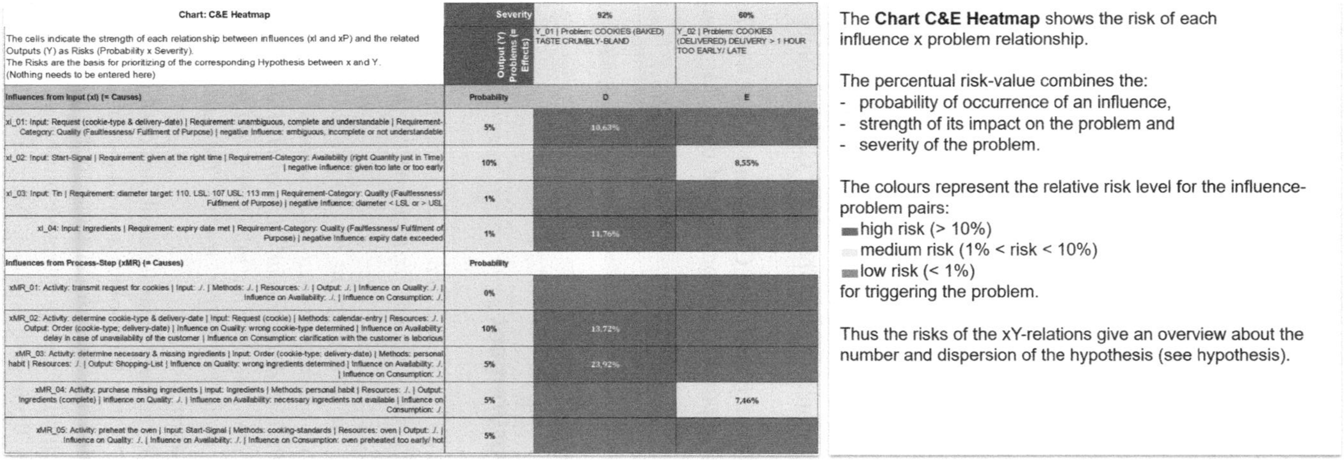

Fig. 7.14 The Chart C&E Heatmap gives an overview about number and dispersion of risks resulting from the causal relationships between influences (x) and problems (Y)

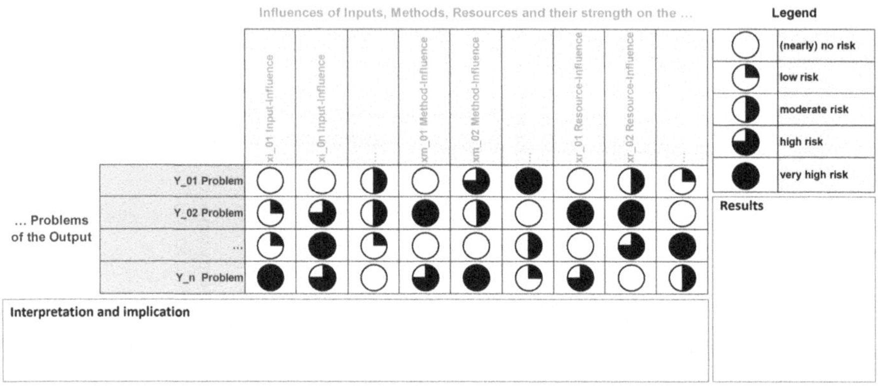

Fig. 7.15 Summary of important influence (x) problem (Y) risks of the C&E Heatmap transformed into 5-level Harvey Ball for a better understanding of your MEASURE phase steering presentation

Tasks

Business-Projects: Please prepare the Cause and Effect Matrix for your project and summarize the most important relationships of the C&E Heatmap in a Harvey Ball table (Fig. 7.15).

Standard-Project: Please prepare the Cause and Effect Matrix for your project and summarize the most important relationships of the C&E Heatmap in a Harvey Ball table (Fig. 7.15).

7.5 Data-Collection[5]

7.5.1 Scales of Measurement

With the preceding tools we have created a model of the reality, with the determined problems and their triggers, with the assumed strengths and the risks of their relationships (Fig. 7.16).

We can now prepare the measurement of x and Y to take samples and then to statistically test hypothesis about their real relationships. This is done in the linked tools: Data-Collection-Plan and the formulation of Hypotheses. Here in the Data-Collection-Plan we will first translate the influences and problems into measurable variables x and Y on their respective scales. This translation is also called Operationalization. Let us go.

You remember that we can only measure attributes of objects because they are the source of variation. This is true for cookies as well as for the loudness of my voice. And it is also true for influences like my distance to the microphone.

Objects can have qualitative and quantitative attributes. And the actual status of these attributes changes from case to case. Measuring now means assigning a

[5]Video 24 (https://www.youtube.com/channel/UCShsy50l5dFsoSXU__hBKSA).

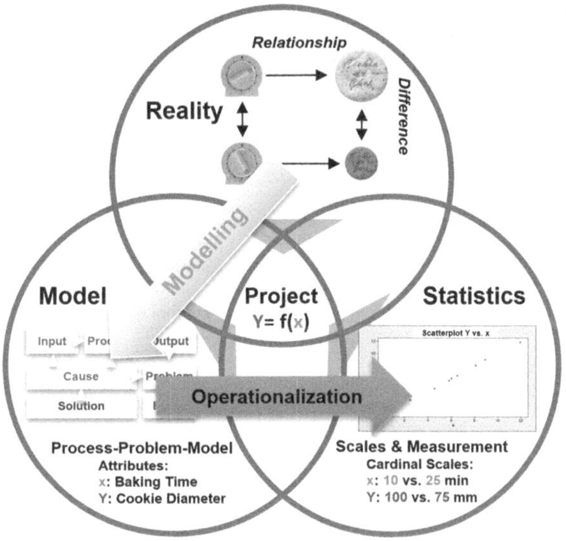

Fig. 7.16 Operationalization of the modeled influences (x) and problems (Y) as measurable variables

Operationalization is the mapping of an object's attribute to a scale,
so that the measured data represent the attribute validly, reliably and precisely.

qualitative or quantitative value of an underlying scale to the respective state of the attribute of an object. With the measurement we want to represent the variation of this attribute as valid and precise as possible by a corresponding variation of the values on a scale. Let us first consider only the individual case, one output, because that is the core of our measurement (Fig. 7.17).

We measure daily, we are informed about measurements, and we are measured ourselves without always being aware of it. We categorize our conversation partner as *male* or *female* and choose one of our available *means of transport*. We award *five stars* for the evaluation of a product and feel personally *good*, *indifferent*, or *bad*. We see the *number* of mails in the inbox and determine our *punctuality* to an appointment.

We could now systematically measure the states of these individual objects for a while. To do this we first need suitable scales whose values reflect the variation of the attributes.

The gender of the people is a qualitative attribute. To measure the gender of our conversation partners we need a quite simple scale with two values: male and female. Even if it does not look like it, but these two values— male/female—form a scale: the **nominal scale**. This also applies, e.g., to the five different values: on foot/bicycle/car/bus and plane, which I have chosen for the scale of the means of

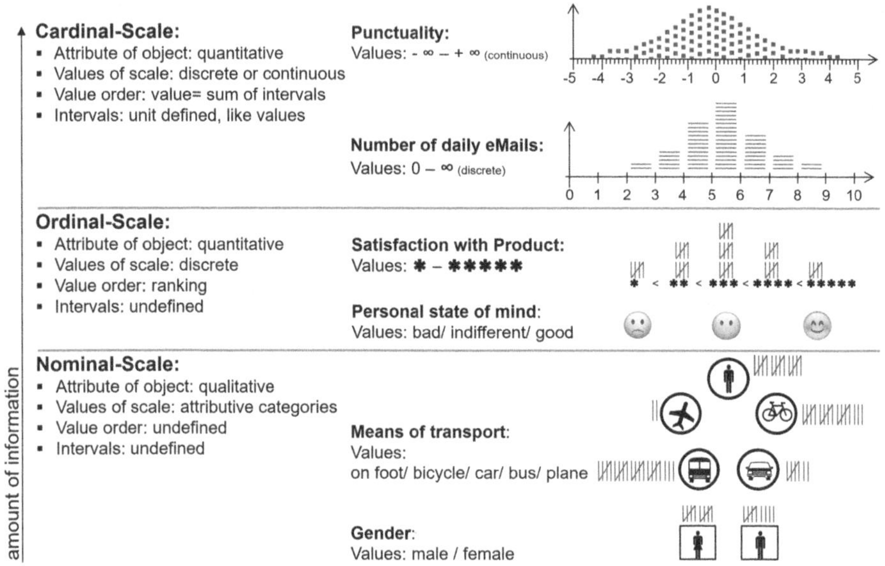

Fig. 7.17 Measuring means assigning a value of an underlying scale to the state of an object's attribute

transport. The values of these two and all other nominal scales have attributive categories as values and not numbers. It follows that there is also no order of values on a nominal scale. And you can see that the order: male/female is just as good and suitable as the order: female/male.

The **ordinal scale**, on the next level, can represent ordered quantitative attributes. In practice it is nearly always realized as a rating scale. Satisfaction assessments and school grades are characterized by the fact that the values and their order are determined on the scale, but not the intervals between the values. We do not really know if a ranking of two stars expresses exactly twice as much satisfaction as one star, because we do not know whether the difference between one star and two stars is equivalent to the difference between zero and one star.

On the upper level, the **cardinal scale**, we can reliably say that 100 emails are twice as much as 50 emails and that 2 min are twice as much as 1 min. The values can be discrete, as with the mails or continuous, as with the time. The units of the intervals between the values are uniformly defined and can be discrete or continuous. The order of values is determined by the principle that each value represents the sum of the preceding intervals.

These attributes of the scales determine the degree of information they can carry. Due to the increasing level of the attributes, the degree of discrimination between the values of the scale increases and thus the amount of information. This means that the values of a sample and thus the underlying attribute can be better discriminated from level to level.

We could collect a sample now for a month and run a tally sheet, e.g., and then we would statistically evaluate the sampled data. Statistics are used to represent a sample graphically to summarize it by parameters and to derive statements about the population. But which graphics are suitable and which parameters may we calculate? (Fig. 7.18)

The most important statistical parameters indicate the (a) central tendency of a sample, like the mean value and (b) its dispersion, such as the standard deviation. But what is the average of 10 women and 9 men? 9.5? Men? Women? Nonsense, that does not work. And the visualization in a Histogram does not work either.

You probably already recognized that the mean cannot be calculated for values on nominal scales and not the standard deviation. Therefore, we also cannot create a Histogram with nominal scaled data. With the time data on punctuality, both would be fine. But we do not have to guess. For each of the three scales the appropriate parameters are defined:

- On the **nominal scale**, the **mode** can be calculated as position parameter, which is female in the gender example. For each election, the winner is the mode. Scattering parameters on the nominal scale are irrelevant to us.
- The position parameter of the **ordinal scale** is the **median** and scattering parameters are, e.g., the **range** and **percentiles**.
- On the uppermost level, the **mean** value and the **standard deviation** can be calculated for the **cardinal scale**.

In addition to the parameters, the scale level also determines the possible graphs and limits the scope of suitable statistical tests. We will see about that later.

The better discriminatory ability from level to level can be seen in the information provided by the parameters because, e.g., the mean carries more information than the median and the median more than the mode. And the same applies to the corresponding scattering parameters.

We have already seen above that the mean value cannot be calculated for nominally scaled data. However, the median—ordinal—could be calculated for the number of daily emails—cardinal. In general: For each selected level the associated parameters and the parameters of the lower level(s) can be calculated. This does not apply the other way around. The parameters of a superior level may not be calculated for data on the selected level. You can easily convince yourself by comparing the formulas and the information they process. For example, neither the mean value nor the standard deviation should be calculated for satisfaction data because on the ordinal scale the intervals are not defined. Nevertheless, this is often done, and we will do it later.

I have added a few more examples from the work environment, everyday life, and for the bakery. Frequently used are, e.g., the defect types at nominal scale level: Cookie burnt, salty, too small and the cycle time or delivery time of the cookies on the cardinal scale (Fig. 7.19).

Make sure you choose the highest possible scale for your measurements. I sometimes saw that, e.g., the measured minutes on punctuality were assigned to

Type of attribute of the object of measurement	Attributes of the Scale	Name of Scale	Parameters of central tendency	Parameters of scattering	Example of everyday life	Example from work environment	Example for cookies
quantitative	Values: continuous Value order: value= sum of intervals Intervals: unit defined, continuous	**Cardinal Scale**	Mean (Xbar)	Average Deviation, Standard Deviation, Variance	race duration: hours : minutes : ...	length: cm cycle time: hours costs: $/ €/ £/ ¥/ ...	weight of cookies: - gram (g) processing time of cookies: - minutes (min)
	Values: discrete Value order: value= sum of intervals Intervals: unit defined, discrete				number of children in a family: 0 ... n	Number of phone calls: 0 ... n	number of defect cookies: 0 ... n
	Values: discrete Value order: ranking Intervals: undefined	**Ordinal Scale**	Median	Range, Percentiles, Interquartile Range	satisfaction assessment: **** smileys grades: A B C D E F / 1 2 3 4 5 6	FMEA Risk-Priority-Number (RPN): 1 - 1000 complexity of projects: project X > project Y	Risk of a Problem: 0% - 100% rating of taste: ****
qualitative	Values: attributive categories Value order: undefined Intervals: undefined	**Nominal Scale**	Mode	number of different values in sample/ number of different possible values in population	election: means of transportation/ parties/ persons/ ...	types of different: - products (Y) - errors (Y) - locations (x) - shifts (x)	types of errors: - burnt - salty - ...
					physical condition: healthy/ sick gender: male/ female	avaliability of a method (x) for certain activity (yes/ no)	Recipe available and used? (yes/ no)

Fig. 7.18 Attributes of objects mapped on scales with different attributes and parameters. The higher the scale level of a measurement, the more information is carried by the data

The **Scale** of a measurand determines the:
- Parameter of central tendency
- Parameter of scattering
- Appropriate statistical test

Values of a higher scale-level can always be transformed to a lower scale level, e.g. if you transform the real age of people (Cardinal Scale) into age categories (Nominal Scale).

This reduction of scale-level reduces the already available information, because the parameters of a Nominal-Scale (e.g. Mode) carry less information then the parameters of a Cardinal-Scale (e.g. Mean).

The same is true for the related statistical tests, where e.g. the Chi^2-Test has less power than the ANOVA.

Thus always plan your measurement on the highest possible level and avoid transformations to a lower scale as far as possible.

Data on a lower scale-level cannot be transformed to a higher scale level

Fig. 7.18 (continued)

↓	C1	C2	C3	C4-T	C5	C6	C7	C8	C9-T
	Y_Cookie-Taste-Chocolate_1	Y_Cookie-Taste-Vanilla_1	Y_Cookie-Taste_1	x_Cookie-Type_		Y_Cookie_Weight_Choc	Y_Cookie_Weight_Van	Y_Cookie_Weight	x_Cookie_Type
1	1	6	1	Y_Cookie-Taste-Chocolate_1		12,0000	10,3101	12,0000	Y_Cookie_Weight_Choc
2	2	7	2	Y_Cookie-Taste-Chocolate_1		5,0000	9,7683	5,0000	Y_Cookie_Weight_Choc
3	3	8	3	Y_Cookie-Taste-Chocolate_1		7,7369	10,1804	7,7369	Y_Cookie_Weight_Choc
4	4	9	4	Y_Cookie-Taste-Chocolate_1		8,4283	9,7450	8,4283	Y_Cookie_Weight_Choc
5	5	10	5	Y_Cookie-Taste-Chocolate_1		7,2708	10,3983	7,2708	Y_Cookie_Weight_Choc
6			6	Y_Cookie-Taste-Vanilla_1		7,6545	10,3738	7,6545	Y_Cookie_Weight_Choc
7			7	Y_Cookie-Taste-Vanilla_1		7,1229	9,6597	7,1229	Y_Cookie_Weight_Choc
8			8	Y_Cookie-Taste-Vanilla_1		8,1624	9,5217	8,1624	Y_Cookie_Weight_Choc
9			9	Y_Cookie-Taste-Vanilla_1		8,1110	9,8849	8,1110	Y_Cookie_Weight_Choc
10			10	Y_Cookie-Taste-Vanilla_1		7,7133	9,8922	7,7133	Y_Cookie_Weight_Choc
11						9,2734	10,4897	9,2734	Y_Cookie_Weight_Choc
12						9,1898	10,2265	9,1898	Y_Cookie_Weight_Choc
13						8,2117	9,5830	8,2117	Y_Cookie_Weight_Choc
14						7,8533	9,6824	7,8533	Y_Cookie_Weight_Choc
15						7,2565	10,0110	7,2565	Y_Cookie_Weight_Choc

Fig. 7.19 Arrangement of sampled data in a Minitab worksheet with individual values for different subgroups (5, 10, and 12) for each variable (C1–C9T). Do not accept aggregated data from reports

categories: too early, on time, too late. By transforming the cardinal scaled minutes data to nominal scaled records, we are giving away information that we later cannot reconstruct from the categories, if required. But of course, we are free to categorize the measured minutes of a sample after we recorded them, if desired (see Fig. 8.14).

In Six Sigma, we also never accept aggregated data such as percentages or means that are already available in reports because for statistics we always need random samples with single raw data.

With the desire to gain as much information as possible about our influences and problems with the measurements of x and Y, we can now move on to the Data-Collection-Plan of sigmaGuide.

7.5.2 Data-Collection-Plan

In the Data-Collection-Plan all outputs and their problems as well as all influences from the inputs and activities are listed. The problems are ranked according to their severity and the influences according of their probability of occurrence (Fig. 7.20).

1. To simplify the translation process from a problem or influence on its measurement, the first question is: **What should be measured?** Please describe here in everyday language what you want to measure to capture the selected problem or the focused influence. For Y_01, for example, we want to measure the *taste of cookies*.
2. Next, we identify the appropriate **measurand**. We were able to measure our customers' brain activity with an EEG while they were eating a cookie, but this would be very expensive and perhaps not valid at all. It would be easier to ask the customer about the taste and have the answers given on a rating scale.
3. Our measurement here will be based on a rating scale with the range of 1 (best) to 6 (worst). And this is the **unit** of measurement, like days, seconds, grams, centimeter, or inch.
4. Now the **target and specification limit** can be specified. We want that our cookies are rated close to very good—1.5—and the lower specification limit is a grade of 4 (Fig. 7.21).

Data Collection Plan		please answer the questions and plan the graphical display as well as initial analyses based on the proposals (right ->)				
Ranking of Influences (xl & xMR) and Outputs (Y)	Influences (xl & xMR) and Outputs (Y)	What should be measured?	Please specify the measurand (e.g. Time)	Please specify the units of the measurand (e.g. days)	Please specify the Target and its Specification Limits - if known - in the format: Target: USL: LSL:	Which different values can the Measurand take? (Scale of Data)
Output (Y)						
1	Y_01 | Problem: COOKIES (BAKED) TASTE CRUMBLY-BLAND	taste of Cookies	rating-scale	grades 1 .. 6	Target: 1.5 LSL: 4	Data Rank Ordered (Ordinal-Scale)
2	Y_02 | Problem: COOKIES (DELIVERED) DELIVERY > 1 HOUR TOO EARLY/ LATE	time intervall between the desired date and the date of delivery	time	hrs	Target: 0 hrs LSL: - 1 hrs USL: + 1 hrs	Data discrete or continuous (Cardinal-Scale)
3	Y_03 | Problem: COOKIES (BOXED) ENERGY-CONSUMPTION > 10% WASTE	energy difference between model calculation and real consumption (=waste of energy)	amount of electricity	kw/h	Target: 0 kw/h USL: + 1 kw/h	Data discrete or continuous (Cardinal-Scale)

Fig. 7.20 Operationalize the measurement of the variables *x* and *Y*, for single case raw data on the highest available scale level

How should the Data be collected?	Is a Measurement-System-Analysis (MSA) necessary?	Which Data about the circumstances should additionally be collected? (Blocking-/ Condition-Variables)	How large should the Sample Size be?	Where should the Data be collected? (Location/ Source)	For which Time Interval should the Data be collected? (Start/ End)	Which Variable-Name will you assign to the Measurand?	In which File will the Data be stored?	Who is responsible for the collection of the data?
collect new data	no	cookie-type (Vanilla vs. Chocolate)	50	shop	1 month	Y_01_Taste	Cookie_du_Chef.xlsx	Green Belt
collect new data	no	Time/ Date/ Supplier	50	bakery	1 month	x_Delay_Desired-Date_Delivery-Date	Cookie_du_Chef.xlsx	Green Belt
collect new data	no	Time/ Date/ Shift	50	bakery	1 month	Y_Waste_Energy	Cookie_du_Chef.xlsx	Green Belt

Fig.7.21 Determine the collection and organization of data

5. The next question—Which different values can the measurand take?—serves to determine the **scale level** of the measurand. The level of a rating scale is ordinal. Optional in this list are additionally: The nominal scale with data in two different categories, such as gender, the nominal scale with data in more than two categories, such as the means of transportation and the cardinal scale, for example, data of time, weight, or size.
6. **Collection**: If we are lucky, we can access already collected data in a database. We do not accept aggregated reports because we need the raw data of individual cases for our sample. In our case there is no data available on our problem and therefore, we need to collect it.
7. The next question concerns the quality of the planned measurement procedure. If there are doubts about the precision, stability, repeatability, or reproducibility of a measuring system and its application, then a **measurement system analysis** (MSA) should be carried out. We do not do this here because we assume that our customers can validly and reliably assess the taste of cookies on a rating scale.
8. The next questions relate to the context of the measurement. A good advice for a further differentiation—later in the statistical tests—is to add some **blocking or condition variables** about the circumstances of your measurement. Time and date of measurement, the supplier or batch number, the shift, the product, and product line are examples of condition variables. I, e.g., choose: Cookie-type as blocking variable. This means that we will measure the taste of vanilla and chocolate cookies to be able to later differentiate the ratings between these cookie types.
9. At this point in time, we can only determine the question of **sample size** by a rule of thumb. The necessary sample size depends on the one hand on the chosen statistical test. Tests that require cardinal scales, such as the regression analysis, have a higher power than tests for nominally scaled data, such as the Chi^2-test and would need a smaller sample size. This topic is in statistics handled under the heading: Power and sample size, and can, if desired, be calculated with Minitab for the common tests. But we do not even know the appropriate tests here yet and we will only get to know them in the next chapter. As a rule of thumb, I enter 50 as the sample size. As a minimum number often, $N = 30$ is suggested because, e.g., correlations $>r = 0.5$ from a sample size $N = 30$ become significant and lower correlations are often irrelevant to us. But I also had to deal with extremely differentiated problems and influences, for which $N = 10$ was sufficient as sample size.

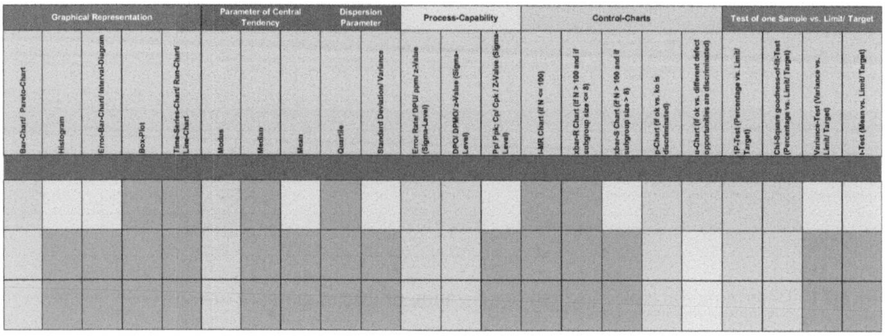

Fig. 7.22 Recommended representations and calculations for each of the operationalized variables x and Y as a first step to an analysis strategy. Green: recommended, yellow: under reservation, grey: not possible (without recoding)

10. In the remaining fields you can now further specify certain information about the **organization of your measurement**, about the location and time interval of your measurement, determine the name of the variable, the file name for its storage, and the responsible person.

After determining the scale level sigmaGuide automatically derives appropriate charts for graphical representations of the data, the appropriate statistical parameters, procedures to calculate the process capability, appropriate Control-Charts and one-sample tests (Fig. 7.22).

- The **green** cells mean full recommendation because the underlying scale carries all necessary information.
- **Yellow** means that you might interpret the results with reservation because the scale does not completely carry the necessary information. The calculation of the mean and standard deviation for ordinal rating scales are only two examples that are widely used in practice.

These suggestions in green and yellow determine the first framework of your analysis strategy. Now you know exactly how you can analyze your data in the first step.

Beside the measurement of our primary problem, I entered additional operationalizations for the other problems and specified influences of the inputs, like the diameter of the received tins and the expiry date of ingredients.

Sometimes it is difficult for me to find a suitable measurand. And in practice, sometimes only an indirect indicator of an attribute can be measured economically, such as the judgment of the cookie's taste. In other cases, I can choose between several operationalizations. If I want to know if the standard requirements of a method are being met, I can (a) register any deviation from the recipe, such as deviating ingredient quantities on a cardinal scale or (b) simply measure whether a recipe has been used or not, at nominal level. I then decide such questions based on how important a high differentiation of the attribute is and what effort is associated with it. In this case, to the usage of a method, a simple yes or no is enough for me.

I think the work on the Data-Collection-Plan is very demanding. Please ask yourself about each of your operationalizations: Can I measure this on an individual object? When you are done, we can move on to the Hypotheses. And here is the good news: sigmaGuide creates the hypotheses automatically.

Questions (SC = Single Choice; MC = Multiple Choice)

Video 24/ Ch. 7.5/ #1 (SC) **Please assign the following measurands to their underlying scale levels.**

		nominal	ordinal	cardinal
1	Cycle time			
2	Different types of waste			
3	Weight in grams			
4	Variants of a product			
5	School grades			
6	Length in millimetres			
7	Defect types			
8	Costs			
9	Satisfaction levels			
10	Human littering types			
10	points			

Video 24/ Ch. 7.5/ #2 (MC) **Please assign the following parameters to the correct scale levels and determine the respective parameter types.**

		nominal	ordinal	cardinal	Position Parameter	Scattering Parameter
1	Mean					
2	Percentile					
3	Range					
4	Mode					
5	Median					
6	Standard deviation					
7	Interquartile range					
8	Variance					
8	points					

Tasks Business-Projects: Please prepare the Data-Collection-Plan for your project.

Standard-Project: Please prepare the Data-Collection-Plan for your project. For your orientation some entries are already prefilled which you can customize according to your needs. To collect data you could observe citizens and assign the xs and Ys according to their behavior.

The questionnaire (see Fig. 7.23) might help you to record the type and frequency of littering (Y), to find triggering causes (x) and to determine the respective littering-type (see Table 6.1) as a root cause (x') later. It can deliver evaluations of the infrastructure of existing trash cans (x) and the perceived cleanliness of the city. And it allows the respondents develop solution ideas (S). The demographic data for

No	Question	Answer						Var
a.	**Actual behavior**							
Q1.	Have you ever thrown away any garbage, such as cans or packaging, in public places?		☐ no			☐ yes		Y
Q2.	Would you tell me how often you throw away garbage in public?	☐ almost never	☐	☐	☐	☐	☐ very often	Y
Q3.	What would you most likely throw away in public? Please rank these objects in a hierarchy. (1= least harmfull to 6= most harmfull)	☐ Metal (cans/ srew caps)	☐ Plastic (bottles/ cups/bags)	☐ Paper (flyer/ newspaper	☐ Cigarette butts	☐ Chewing gum	☐ Glas (bottles/ jars)	Y
	Which of these statements apply most to you?							
Q4a	When I'm pressed for time, I sometimes throw garbage away	☐ disagree	☐	☐	☐	☐	☐ agree	X
Q4b	Sometimes I put garbage in a place where it cannot be seen	☐ disagree	☐	☐	☐	☐	☐ agree	X
Q4c	In places where there is already garbage lying around, it is not so bad to add something to it	☐ disagree	☐	☐	☐	☐	☐ agree	X
Q4d	When my friends throw something away, I join them sometimes	☐ disagree	☐	☐	☐	☐	☐ agree	X
Q4e	I don't mind the garbage on the ground. And a bit more or less doesn't matter.	☐ disagree	☐	☐	☐	☐	☐ agree	X
Q4f	Throwing garbage away is not a problem. The garbage removal service is paid to clean it up.	☐ disagree	☐	☐	☐	☐	☐ agree	X
Q4g	I sometimes like to provoke other people by throwing my garbage on the ground	☐ disagree	☐	☐	☐	☐	☐ agree	X
Q5.	Have you changed the extent of littering in recent years?	☐ extremely decreased	☐ clearly decreased	☐ slightly decreased	☐ slightly increased	☐ clearly increased	☐ extremely increased	X
b.	**Support from our city**							**b.**
Q6.	How well is our city equipped with trash cans?	☐ very bad	☐	☐	☐	☐	☐ very good	X
Q7.	How clean are our streets and squares in your opinion?	☐ very dirty	☐	☐	☐	☐	☐ very clean	X
c.	**Solution**							**c.**
Q8.	What should our city do to prevent the litter?							S
Q9.	What would have to happen in order for you to improve your behaviour towards the environment (even further)?							S
d.	**Sociodemgraphic information about the interviewpartner (estimations)**							**d.**
Q10.	Gender		☐ female			☐ male		X
Q11.	Age (estimated)	☐ < 20	☐ 20-30	☐ 30-40	☐ 40-50	☐ 50-60	☐ > 60	X
Q12.	Economic Status (estimated)	☐ low		☐ normal		☐ high		X
Location:		**Date:**		**Time:**		**No:**		

Fig. 7.23 Mandatory questionnaire for interviews on littering behavior of citizens (download)

Table 7.1 Plan for a stratified sample ($N = 40$) for conducting the interviews (mandatory)

	female	male	female	male
	< 30 years old		>30 years old	
busstop	5	5	5	5
park	5	5	5	5

each respondent can be regarded as separate influences (x) and can serve as condition/blocking variable for the evaluation of problems (Y) (see Fig. 7.21).

To support your data collection, please use the prepared questionnaire (available as download) and supplement it if necessary, according to your requirements. A questionnaire in a Six Sigma project is not self-evident, but also not unusual and it is binding for this project. The following plan can help you organize your data collection and can supplement other data that you collect.

A stratified sampling plan helps to reduce the sampling error. It has the advantage that you can later evaluate the entire data factorially, e.g., to compare men with women, younger with older people and people who are on the move with people who are relaxing. This makes the factors influence variables (x) and you can statistically test the differences between the factor levels (Table 7.1).

There are two risks that you will not get honest answers in your interviews:

(a) **Social desirability:** People like to present themselves better than they are. So that the answers are not polished, you need the trust of the people. Never show them a negative attitude toward their answers. Always be polite and thank them for every answer.
(b) **Trash blindness:** Some people do not notice the garbage lying around. They do not see it until you point it out to them. Therefore, look for a place for your interviews where some garbage lies around.

Additionally, we require collecting data about the actual littering at 20 different places near trash cans. You probably know places in your hometown that are often overcrowded or often deserted (Table 7.2).

The task is to go to these selected places and the count or at least estimate and classify the garbage according to the categories of the questionnaire (metal, plastic, paper, cigarette butts, chewing gum, glass).

Please adapt your data collection plan in sigmaGuide to these specifications. Of course, you can also develop your own ideas for data collection and adapt your Data-Collection-Plan accordingly. It is also not mandatory for you to collect the data at parks and bus stops if you want to compare different places. In order to perform a

Table 7.2 Plan for a stratified sample ($N = 20$) of your field study (mandatory)

	rest places (e.g. park) near trash can	traffic places (e.g. busstop) near trash can
often overcrowded	5 locations; count and classify the garbage in a radius of 5 m around the trash can	5 locations; count and classify the garbage in a radius of 5 m around the trash can
often deserted	5 locations; count and classify the garbage in a radius of 5 m around the trash can	5 locations; count and classify the garbage in a radius of 5 m around the trash can

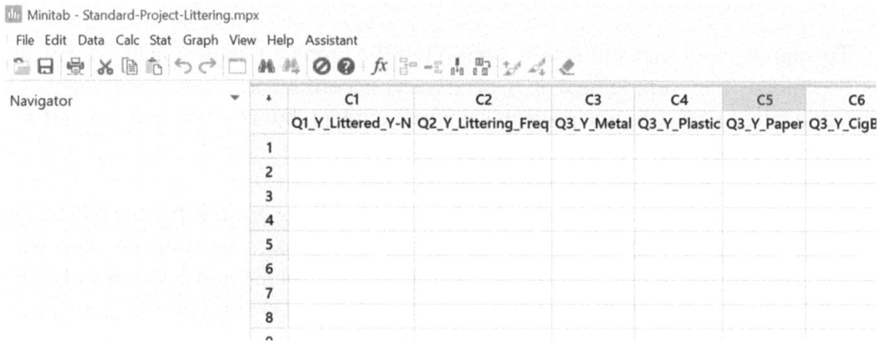

Fig. 7.24 Minitab worksheet for collected data of the interviews and the field study

statistical evaluation, e.g., with a Chi^2-test, the subgroup size in each cell of your collection plans must be at least $n = 5$.

Your worksheet in Minitab for the data of your interviews and your field study might then look like shown in Fig. 7.24 (download).

7.6 Hypotheses[6]

We have already defined the problems—the Ys, identified their triggering causes—the xs and evaluated risks of x–Y pairs, based on expert assessments. Now we need hypotheses if we want to empirically analyze the measurement data of these x–Y pairs (Fig. 7.25).

[6]Video 25 (https://www.youtube.com/channel/UCShsy50l5dFsoSXU__hBKSA).

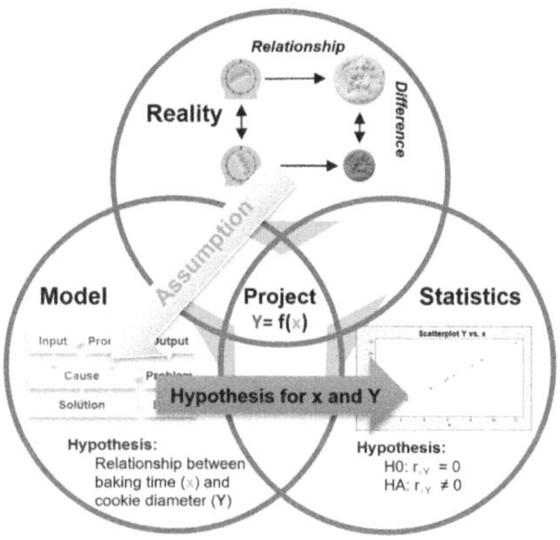

Fig. 7.25 Observations of the reality are transformed into logical assumptions and then translated into hypotheses about relationships or differences

A **hypothesis** is the translation of an assumption about objects in reality into an empirically testable format with the measurands x and **Y**.

7.6.1 Modeled Content of Assumptions: Difference Hypothesis vs. Relationship Hypothesis

Causality explains life and makes it predictable. We observe events in reality and then we generalize our findings from the individual cases to assumptions about causality, e.g., in the statement:

- *If it rains (x), then the road is wet (Y).*

The inversion of this *If–Then* statement leads to the: *Why? Because!* statement: *Why is the road wet (Y)? Because it is raining (x)!*

This is the core of the Root-Cause-Analysis (5 Whys) that we will apply at the end of the analysis phase. Logically, however, both statements are identical.

A more precise formulation of the categorial statement: *If x, then Y* is the assumption:

- *The more it rains (x), the wetter the road (Y)*

with a continuous relationship between *x* and *Y*. And in a final statement we can formulate: *Y differs between the states of x* which leads to the assumption, for example:

- *The amount of wetness (Y) differs from road to road (x).*

Relationship-Hypothesis:

- **Assumption**: If x, then Y
- **Attributes of**
  x: 2 … n levels (categories on nominal scale)
  Y: 2 … n levels (categories on nominal scale)
- **Hypothesis**:
  There is a relationship between gender (x) and the type of preferred snack (Y)
- **Test**: e.g. Chi²-Test

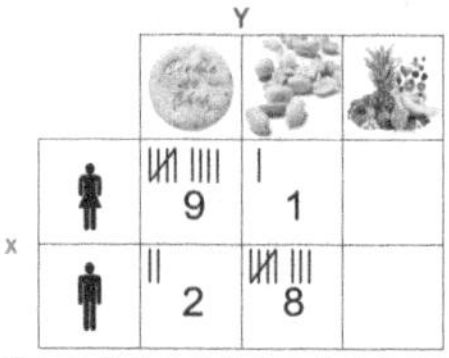

Relationship-Hypothesis:

- **Assumption**: The x, the Y
- **Attributes of**
  x: discrete or continuous data (ordinal or cardinal scale)
  Y: discrete or continuous data (ordinal or cardinal scale)
- **Hypothesis**:
  There is a relationship between the baking time (x) and cookie diameter (Y)
- **Test**: e.g. Correlation, Regression

C126	C127
x_bakingTime	Y_cookieDiameter_
12	9,8333
9	10,1111
18	9,5556
25	9,4000
10	10,0000
13	9,7692

Difference-Hypothesis:

- **Assumption**: Difference in Y, dependent from type/ level of (x)
- **Attributes of**
  x: 2 … n levels (categories on nominal scale)
  Y: discrete or continuous data (ordinal or cardinal scale)
- **Hypothesis**:
  There is a difference in the weight of cookies (Y) between cookie types (x)
- **Test**: e.g. t-Test, ANOVA

C129-T	C130
x_cookieType	Y_cookieDiameter
Vanilla	10,5
Vanilla	10,2
Chocolate	8,7
Vanilla	9,9
Chocolate	13,8
Chocolate	11,1

Fig. 7.26 Observations of the reality are transformed into logical assumptions and then translated into hypotheses about relationships or differences

We will first translate these three assumptions into two types of hypotheses expressing the content of the observation and then translate them further into the statistical format of H0 and HA hypotheses (Fig. 7.26).

The advantage of this approach is that we can determine the appropriate statistical tests simultaneously with the type of hypothesis. And the type of hypothesis is determined by the scale level of *x* and *Y*. For example, we could have observed, that: Men prefer salty peanuts over TV and women prefer sweet cookies leading to the assumption: If male (*x*), then preference for peanuts (*Y*), if female (*x*) then preference for cookies (*Y*).

These statements can of course be seen as prejudices, as well as the many prejudices that people express based on individual observations. But for us these are assumptions, that we want to check with hypotheses, based on collected data in a corresponding table (see Figs. 7.19 and 7.24).

We could also expand this example and add the category: *fruit* to the TV snack and *transgender* as an additional gender category. However, this requires further If–Then assumptions, which become confusing and impractical as the number of categories increases.

7.6.2 Relationship Hypothesis

An efficient representation is provided by the simple format: *There is a relationship between x and Y*. This **relationship hypothesis** is appropriate for at least one *x* (e.g., gender) and one Y (e.g., snack preference), each with attributive categories and thus nominally scaled. And the relationship could be statistically examined with the Chi^2-test.

We cannot apply the more precise continuous version of the If–Then assumption to this example, because the statement: *The gender (x), the preference (Y)* is pointless. In another example we could however assume that the baking time affects the color or the diameter of a cookie: *The longer the baking time (x) the darker or smaller is the cookie (Y).* In these cases, both x and Y are discrete or continuous numbers on ordinal or cardinal scales. However, the format of the hypothesis remains identical to the previous example:

- *There is a relationship between the baking time (x) and cookie diameter (Y).*

We could also collect data for this hypothesis and record it case by case in two columns of a table (see Figs. 7.19 and 7.24). This relationship could be statistically investigated, for example, with the correlation or a simple regression analysis.

Please note that we distinguish the concept of *relationship* from the concept of *correlation* because behind the term relationship we assume a causal relationship between x and Y in the model, while the statistical method of correlation is directionless and does not allow any statement about causality. The causality lies in the assumption of our model, the statistics link numerical measurements to numerical results.

7.6.3 Difference Hypothesis

For the third assumption we must change the wording of the hypothesis. Because the assumption: Difference in Y, dependent from type or level of (x) is based on discrete or continuous data on an ordinal or cardinal scale for Y and 2 to n categories on a nominal scale for x. We could assume and formulate:

- *There is a difference in the weight of cookies (Y) between types of cookies (x).*

This hypothesis can be tested, e.g., by the t-Test or the ANOVA, if the data are entered again as measurement pairs in two columns of a table (see Figs. 7.19 and 7.24).

7.6.4 Statistical Format of Assumptions: Null Hypothesis (H0) vs. Alternative Hypothesis (HA)

But how can we translate these hypotheses, which are based on causal assumptions in the model, into the well-known format of a statistical hypothesis? (Fig. 7.27)

That is very easy. In the *Null hypothesis* (H0) we simply formulate that there **is *no* *relationship*** or ***no** difference*. And in the *Alternative hypothesis* (HA) we formulate

... and translated into the format of statistically testable hypotheses

Assumption about reality modeled as ...	Null-Hypothesis (H0)	Alternative-Hypothesis (HA)
Relationship-Hypothesis	From the assumption about reality...	
	There is no relationship between x and Y	There is a relationship between x and Y
	Example	
	There is a/no relationship between baking time (x) and cookie size (Y)	
	... to the statistically testable formulation	
	$H_0: r_{xy} = 0$	$H_A: r_{xy} \neq 0$
Difference-Hypothesis	From the assumption about reality...	
	There is no difference - in the degree of: Y - between the levels of: x (xi, xj, ...)	There is a difference - in the degree of: Y - between the levels of: x (xi, xj, ...)
	Example	
	There is a/ no difference - in: the taste of cookies (Y) - between: types of cookies (xi) (e.g. Vanilla (x1) vs. Chocolate (x2) vs. ...)	
	... to the statistically testable formulation	
	$H_0: Y_{xi} = Y_{xj}$	$H_A: Y_{xi} \neq Y_{xj}$

In reality, we observe relationships and differences that lead us to assumptions of causal connections between x and Y.

These assumptions about reality are represented in the model as hypothesis about:

- **Relationships** between x and Y or
- **Differences** in Y between levels of x.

based on the scale level of x and Y.

These relationship and difference hypotheses can be statistically tested when translated into the format of statistical hypotheses. In the:

- **Null-Hypothesis** (H_0) **no** relationship or difference is formulated
- **Alternative-Hypothesis** (H_A) **a** relationship or difference is formulated.

Therefore, the Data-Collection-Plan already defines a) the type of hypothesis as relationship or difference and b) determines the appropriate statistical tests.

The statistical tests now checks the Null-Hypothesis and indicates whether the H_0 is confirmed or should be rejected in favour of H_A, based on the calculated p-value of the test and the alpha-level, that we have determined in advance.

Fig. 7.27 The logical content of each hypothesis on relationships or differences is formally divided into the hypotheses H0 vs. HA for their statistical analysis

that there ***is a** relationship* or that there ***is a** difference*. So, we can simply combine our content hypotheses and the statistical hypotheses in one table. To keep the hypotheses clear, we can link their exclusive statements to one hypothesis:

There is a vs. no relationship ... respectively
There is a vs. no difference ...

In the restrictive statistical formulation, the hypotheses are as follows: The assumption of no relationship corresponds to the null Hypothesis (H0):

- *The relationship between x and y is zero ($rx\text{–}Y = 0$).*

And the assumption of a relationship corresponds to the alternative hypothesis (HA):

- *The relationship between x and y is unequal zero ($rx\text{–}Y \neq 0$).*

The same applies to the difference hypothesis. No difference, null hypothesis (H0):

- *Y under the condition of xi is equal to Y under the condition of xj ($Yxi = Yxj$).*

And we could also say that:

- *The difference between Yxi and Yxj is zero ($Yxi - Yxj = 0$).*

In the alternative hypothesis (HA) we state that:

- *Yxi is unequal to Yxj ($Yxi \neq Yxj$)* or that their *difference is unequal to zero ($Yxi - Yxj \neq 0$).*

And now you can relax because with these preparations, with the prioritization of the risky *x–Y* pairs in the Cause and Effect Matrix and with the determination of the scale level for the measurands of *x* and *Y* in the Data-Collection-Plan, the appropriate hypotheses can be derived, as well as the tests with which the hypotheses can be examined.

7.6.5 Scale Levels Determine Appropriate Statistical Tests

Each statistical test sets conditions for its application. The first condition is the scale level of the variables to be tested. Some tests do not work at all if the scale level does not fit. This means, e.g., that the correlation cannot be applied to nominal-scaled variables (Fig. 7.28).

X \ Y	Data in 2 Levels (Nominal-Scale)	Data in > 2 Levels (Nominal-Scale)	Data Rank Ordered (Ordinal-Scale)	Data discrete or continuous (Cardinal-Scale)
Data in 2 Levels (Nominal-Scale)	Relationship Hypothesis Chi-Square-Test	Relationship Hypothesis Chi-Square-Test	Difference Hypothesis Wilcoxon-Mann-Whitney-Test	Difference Hypothesis t-Test
Data in > 2 Levels (Nominal-Scale)	Relationship Hypothesis Chi-Square-Test	Relationship Hypothesis Chi-Square-Test	Difference Hypothesis Kruskal-Wallis-Test	Difference Hypothesis ANOVA
Data Rank Ordered (Ordinal-Scale)	Relationship Hypothesis Binary-Logistic-Regression	Relationship Hypothesis Nominal-Logistic-Regression	Relationship Hypothesis Rank Correlation (Spearman) / Ordinal-Logistic-Regression	Relationship Hypothesis Rank Correlation (Spearman)
Data discrete or continuous (Cardinal-Scale)	Relationship Hypothesis Binary-Logistic-Regression	Relationship Hypothesis Nominal-Logistic-Regression	Relationship Hypothesis Rank Correlation (Spearman) / Ordinal-Logistic-Regression	Relationship Hypothesis Product-Moment-Correlation (Pearson) / General Regression

The scale level of x and Y determines the content of the hypothesis and the selection of permissible statistical tests.

Based on the determined scale level in the Data-Collection-Plan, sigmaGuide automatically formulates hypotheses for xY pairs and suggests suitable tests.

Fig. 7.28 Selection of the appropriate statistical test for the examination of hypotheses depending on the determined scale level of the involved variables *x* and *Y*

In other cases, tests are used although there are more suitable tests for the existing scale level or that test results may only be interpreted with reservation for the selected test. To make this decision easier for you, I have listed the appropriate hypothesis for simple *x*–*Y* pairs in this table, as well as suitable parametric and non-parametric tests for their evaluation.

In this table you will find the *x* variable and its possible scale levels in the rows and the *Y* variable and its possible scale levels in the columns. The appropriate hypothesis and the suitable tests are then listed at the intersection of rows and columns. So, you can find the relationship hypothesis and the Chi^2-test, if *x* and *Y* are nominally scaled, as in our gender–snack example. The *t*-Test to evaluate a difference hypothesis is suitable, if we want to compare the weight (*Y*) of vanilla and chocolate cookies (*x*). If we want to compare even more types of cookies—vanilla, chocolate, coconut—and thus if *x* is nominally scaled in more than two levels, then the ANOVA is appropriate for this difference hypothesis.

If you want to check whether the baking time (*x*) affects the size of the cookies (*Y*), with two cardinally scaled measurands, then you can test this relationship hypothesis with the correlation or with the simple regression analysis.

These are hypotheses and tests that are very common in practice. But there are of course other variants, which you can then simply read in the table yourself—or you leave the work to sigmaGuide.

7.6.6 Automatic Generation of Hypotheses in sigmaGuide

sigmaGuide uses the determined scale level of the Data-Collection-Plan to formulate the appropriate hypothesis as a relationship or difference and at the same time determines the possible statistical tests for their evaluation. And as a result, it lists all problems (Y), their risks from the x–Y pairs, the type of hypothesis and its concrete formulation as well as the suitable statistical tests. To sort the list by the risk level of the x–Y pairs, use the command: *Ctrl+s* (Fig. 7.29).

The first hypothesis states a difference in the rating of the taste of cookies (Y) between types of cookies (x). The assessed risk of this influence–problem pair is 10.63% and the appropriate tests are the Wilcoxon–Mann–Whitney Test and the *t*-Test. The formulation of the hypotheses is somewhat difficult to read because they also contain information about the variables. But you will recognize your entered information so far when you read it (Fig. 7.30).

You should, however, not force your Sponsor to read this bulky wording. Instead, I recommend that you use the simple formulations of the previous pages for your presentation: *There is a relationship between x and Y* and: *There is a Difference in Y between levels of x* to familiarize your Sponsor with your hypotheses (see Chap. 5).

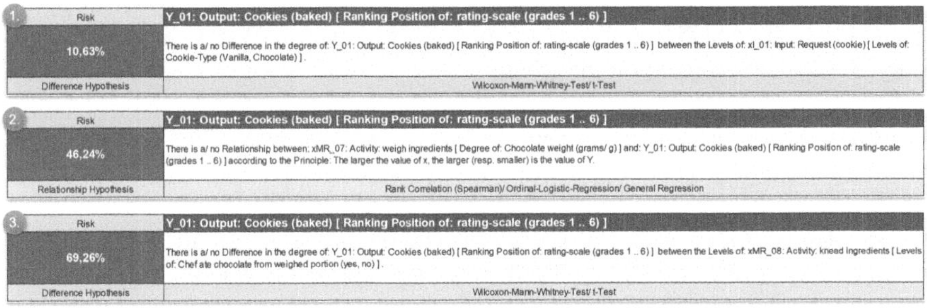

Fig. 7.29 Automatically generated (univariate) hypothesis for Y and xi, xm and xr, their risk to cause a problem and recommended statistical tests

Relationship-Hypotheses					
There is a relationship between: x ...	**... and Y**	**Scale Level of Y**	**Scale Level of x**	**Graphical Representation:**	**Statistical Test:**
Amount of Chocolate in ingredients	*Rating of Taste*	*cardinal*	*cardinal*	*Scatter Plot*	*Correlation/ Regression*
...	...	...	...	...	...

Difference-Hypotheses					
There is a difference in: (Y)	**between: Levels of (x)**	**Scale Level of Y**	**Scale Level of x**	**Graphical Representation:**	**Statistical Test:**
The taste rating of Cookies	*Types of Cookies (Vanilla vs. Chocolate)*	*cardinal*	*nominal*	*Box-Plot*	*t-Test/ ANOVA*
...	...	...	...	...	...

Fig. 7.30 Templates for hypotheses to facilitate formulating additional hypotheses, not mentioned in sigmaGuide

I will focus on the first three hypothesis in the ANALYSIS phase. Let me please finish the MEASURE phase here, because I need some time to collect the data according to my Data-Collection-Plan.

Questions (SC = Single Choice; MC = Multiple Choice)

Video 25/ Ch. 7.6/ #1 (SC)	Which hypothesis is ...		
		H0	HA
1	... tested in a statistical test?		
2	... accepted if p < alpha?		
3	... accepted if p >= alpha?		
3	points		

Video 25/ Ch. 7.6/ #2 (MC)	Which scale levels of x and Y define the type of content hypothesis?				
		x: nominal	x: cardinal	Y: nominal	Y: cardinal
1	Difference Hypothesis				
2	Relationship Hypothesis				
2	points				

Tasks Business-Projects: Please check the hypotheses in sigmaGuide and add further hypotheses according to the templates in Fig. 7.30, if necessary. Copy the results of all MEASURE tools and the pictures from the workshop in your Project-Story-Book and comment on the results.

Standard-Project: Please check the hypotheses in sigmaGuide and add further hypotheses according to the templates in Fig. 7.30, if necessary. The division of the questionnaire into *x*s and *Y*s opens up many possibilities for additional hypotheses, e.g., based on the assumed relationships between Q2 and Q4.

Copy the results of all MEASURE tools and the pictures from the workshop—if available—in your Project-Story-Book and comment on the results.

ANALYSE

8

8.1 Overview: Steps, Tools, Their Purpose, and Results[1]

In the MEASURE phase we modelled our observations of reality as assumptions about causal relationships and differences to x and Y. We have operationalized risky x–Y pairs as measurands, translated them into content hypotheses and converted them into a statistically verifiable format.

The core of the ANALYSIS phase now is to test these hypotheses statistically based on the collected data and then to determine the root causes (x') of the triggering influences (x) of the problems (Y).

The identification of the root causes (x') is the primary purpose of the ANALYSIS phase, because they are the interface for solution ideas in the IMPROVE phase.

Please note that by testing the hypotheses, we can only confirm or reject our assumptions about causal relationships in reality. Statistics, however, do not provide any proof of causality but can merely support the usefulness of our assumptions in the model (Figs. 8.1 and 8.2).

First, we will evaluate the collected data. To do so we check the data set for plausibility and correct or eliminate incorrect entries. In a graphical analysis we then check their distribution and chronological profile.

Next, we will calculate the process capability, e.g., DPMO, Pp/Ppk, and Sigma Level and we will monitor the process performance over time with Control-Charts.

Then we test the hypotheses with the highest risk values statistically with the recommended tests.

We then examine the triggering influences (x) on their root causes (x') and thus identify the interfaces for solutions with which the root causes can be eliminated, adjusted, or circumvented.

---

[1]Video 26 (https://www.youtube.com/channel/UCShsy50l5dFsoSXU__hBKSA).

© Springer Nature Switzerland AG 2019

R. Hutwelker, *Six Sigma Green Belt Certification Project*, Management for Professionals, https://doi.org/10.1007/978-3-030-31915-1_8

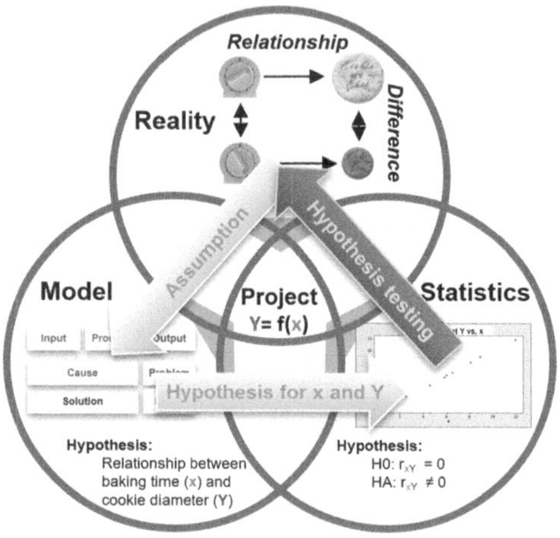

Fig. 8.1 Completion of the investigation cycle: Reality, modeling, and testing

> By testing the hypotheses we can **confirm** or **reject** our modelled **assumptions** about causal relationships in reality.
> However, the **statistics do not provide any proof of causality**, but merely support the usefulness of our assumptions in the model.

The presented procedure will show all important steps of data analysis. By linking the hypothesis tests with the Root-Cause-Analysis, I would also like to introduce an analysis strategy that you can apply to your project.

With the ANALYSIS phase we also finish the scientific part of Six Sigma. The procedure in the IMPROVE phase for the development of solution ideas and implementation of concrete measures corresponds to the typical procedure of other improvement projects.

8.2 Data Evaluation[2]

I first enter all collected data into a spreadsheet program, e.g., Excel, before transferring it into a statistics program, e.g., Minitab. This has the advantage that I can organize, sort, and filter the data very comfortably. In addition, existing data

[2]Video 27 (https://www.youtube.com/channel/UCShsy50l5dFsoSXU__hBKSA).

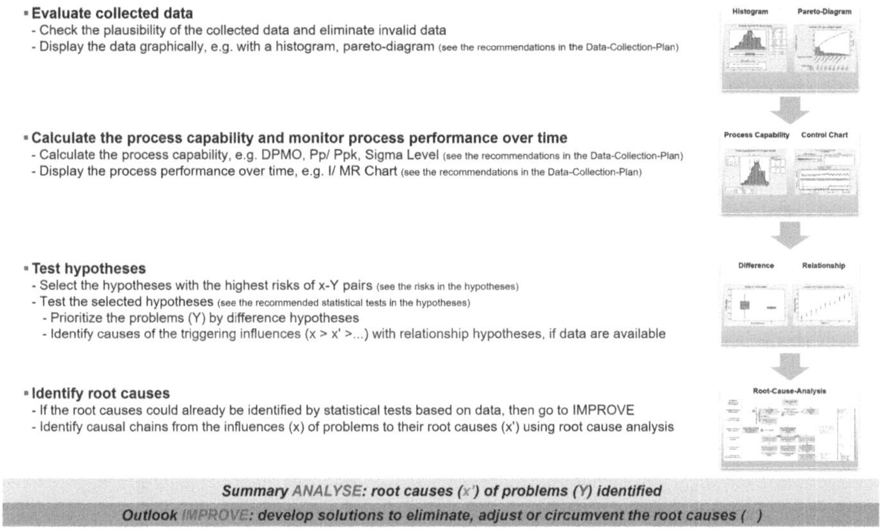

Fig. 8.2 Overview ANALYSE: From the visualization of problems (Y) to the identification of their root causes (x')

from databases can be exported to Excel very easily. And here in Excel I label the columns with meaningful variable names from the Data-Collection-Plan (Fig. 8.3).

Preparation of Data Many users insert additional data rows in Excel to format the table and make it clearer. In the statistics program, however, empty cells are regarded as missing data. Therefore, delete these empty formatting rows before you transfer them to your statistics program.

Minitab issues an error message if columns, linked in the same analysis, do not have the same length. Therefore, delete unnecessary entries at the end of a column.

Transfer of Data to MINITAB Raw data of an Excel spreadsheet can either be copied and pasted directly into a Minitab worksheet, or imported via the menu item: *File/Open*.

If you want to paste copied data, make sure you select the column label (*Cn*) in Minitab and then paste. Otherwise the Excel variable name will be the first data entry in the data row, and you will get the variable name as data. This also means that the data type of this variable is automatically set to: TEXT, which corresponds to our nominal scale level.

Plausibility Check in Minitab Minitab distinguishes three data types: NUMERIC, indexed in the column label in the first cell by: *Cn*, suitable for our ordinal and cardinal scaled data. The data type: TEXT (*Cn-T*) is suitable for our attributive nominal scaled data and the additional data type: DATE/TIME (*Cn-D*) for time specifications.

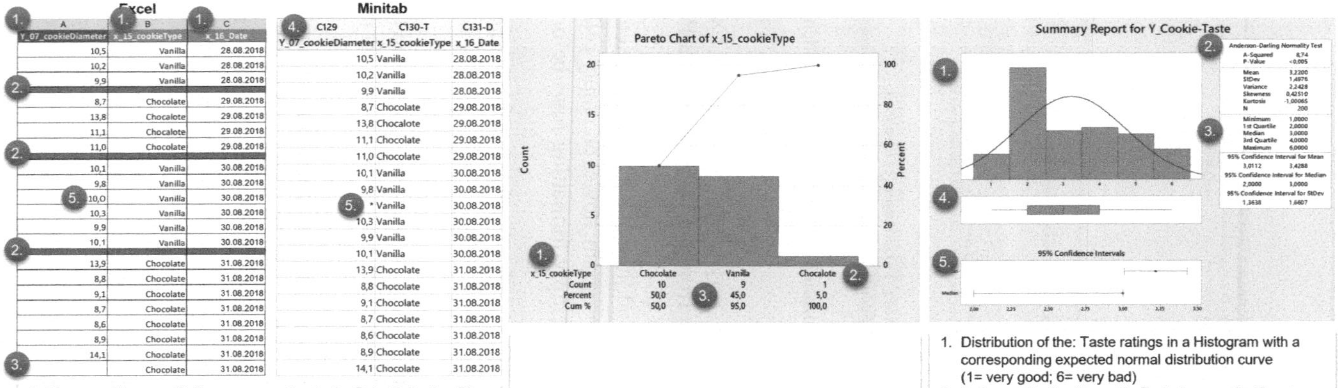

1. Name problems and influences according to the Data-Collection-Plan, with a meaningful name, with a Y in front of the problems and an x in front of the influences, supplemented by index numbers.
2. Delete additional rows you may have inserted to structure the table.
3. Make sure that related data columns have the same length. If necessary, delete unneeded entries at the end of a column.
4. Format the data in the Excel table if, e.g., Minitab uses the **T** to display an attributive variable, although they are numeric data.
5. Correct obvious input errors, like the wrong **O** instead of the correct **zero**, which are indexed in Minitab with an asterisk as missing data.

1. Count, percentage proportions and cumulative percentage proportions of cookie types in the sample
2. Wrong spelling - Chocalote - and therefore input errors. Correct wrong entries in the raw data before you proceed.
3. The frequency of chocolate and vanilla cookies in the sample is approximately the same

1. Distribution of the: Taste ratings in a Histogram with a corresponding expected normal distribution curve (1= very good; 6= very bad)
2. Anderson-Darling Normality Test shows a significant result; (p< 0,005), meaning: data are not normally distributed. Additionally important statistics are shown: Mean, Standard Deviation, Median, …
3. Data are in the expected range, from 1 to 6; data outside this range would be invalid
4. Box-Plot of the Distribution (Histogram seen "from above")
5. Confidence Intervals for the Mean and Median. It means, that with a probability of 95% the Mean/ Median in the population will be within this range.

Fig. 8.3 Input of collected data, correction of invalid and useless data, and plausibility check of the data in the Pareto-Chart and Histogram

If numeric data is formatted as text in Excel, Minitab also displays it in TEXT format and indexes it with a: *T* in the column label. In this case, the data cannot be evaluated numerically. Therefore, numerical data represented in Minitab as a TEXT variable (*Cn-T*) must be reformatted either in Excel or with the command: *Data/ Change Data Type* in the Minitab menu.

Some incorrect entries in Excel, such as an: O instead of a: zero, are indexed by Minitab with an asterisk (∗) as missing data. It is always a good idea to first correct the raw data in Excel and copy the corrected data back into Minitab or import the complete file again.

After data entry and correction of invalid and useless raw data, you can continue the plausibility check with graphical evaluations.

The Pareto-Chart, suitable for nominal scaled data, shows that the sample contains 50% chocolate cookies, 45% vanilla cookies, and 5% *chocolate* cookies. The *chocolate* cookie is again a data entry error, which I correct in the raw data before I continue my inspection.

The Histogram is suitable for cardinal and, with restrictions, for ordinal scaled data. The Anderson–Darling Test shows that this data is not normally distributed for the taste ratings of all cookies. We could now create separate Histograms for chocolate and vanilla cookies to check the normality of the distributions, but that is not the purpose of the data inspection.

To check the plausibility of the data I look at the: Minimum and Maximum values. They are in the expected range from 1 to 6, as determined in the Data-Collection-Plan. Data below and above this range would again indicate invalid data.

I save the complete raw data from Excel in a Minitab worksheet. And I store filtered data from Excel in additional worksheets. Make sure you name all worksheets accordingly to keep track of them.

Questions (SC = Single Choice; MC = Multiple Choice)

Video 27/ Ch. 8.2/ #1 (SC)	Please assign the following diagrams to the lowest necessary scale level.			
		nominal	ordinal	cardinal
1	Histogramm			
2	Box Plot			
3	Time Series Plot			
4	Pareto Chart			
5	Scatter Plot			
6	Interval Plot			
6	points			

Tasks

Business-Projects: Please evaluate your data, e.g., with Minitab, as shown here. Minitab is available in a 30-day trial version. The download link is specific for many countries. When you type: *Download Minitab* into your search engine, you will probably find the corresponding link.

Standard-Project: Please evaluate your data, e.g., with Minitab, as shown here. Minitab is available in a 30-day trial version. The download link is specific for many

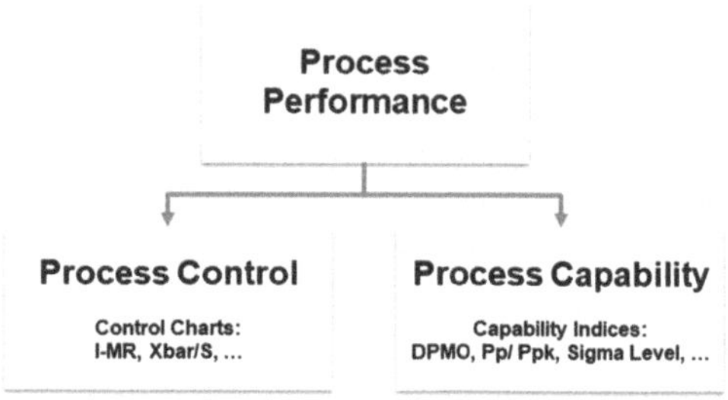

Fig. 8.4 The process performance is determined: (**a**) at a certain point in time (Process Capability) and (**b**) over time (Control Charts)

countries. When you type: *Download Minitab* into your search engine, you will probably find the corresponding link.

8.3 Process Performance[3]

Meanwhile, my sample has increased to N = 100 for chocolate and vanilla cookies. And, of course, I always include all available data, measured under similar conditions and not older than one year, even if the calculated minimum sample size of the Data-Collection-Plan is exceeded. This reduces the confidence intervals and makes the calculations even more reliable.

And later I can compare the data of one year before improvement with the data of one year after improvement. Seasonal fluctuations can then also be considered—but unfortunately this is often only possible with a great deal of effort (Fig. 8.4).

First, I create a graphical summary again with the Minitab Assistant (*Assistant/ Graphical Analysis/Graphical Summary*). Here I divide my sample by cookie type in chocolate vs. vanilla cookies.

The separated Histograms for chocolate and vanilla cookies with the expected overlaid normal distribution curve again indicate:

(a) A deviation from the normal distribution ($p < 0.005$)
(b) Different means (3.98 vs. 2.46) and
(c) Different standard deviations (1.645 vs. 0.796)
(d) Within the expected data range from: 1 = very good to 6 = very bad.

The Time-Series-Plot shows for the vanilla cookies a stationary course of events and for the chocolate cookies a shift in the taste rating from good to bad. This means: There is a difference in taste between the two types of cookies. But is this difference statistically significant and practically relevant? We will test this difference later in the hypothesis tests.

First, I would like to point out that the Time-Series-Plots could be transferred to the Histograms if we divided the *Y*-axis into the value categories of the Histogram

[3]Video 28 (https://www.youtube.com/channel/UCShsy50l5dFsoSXU__hBKSA).

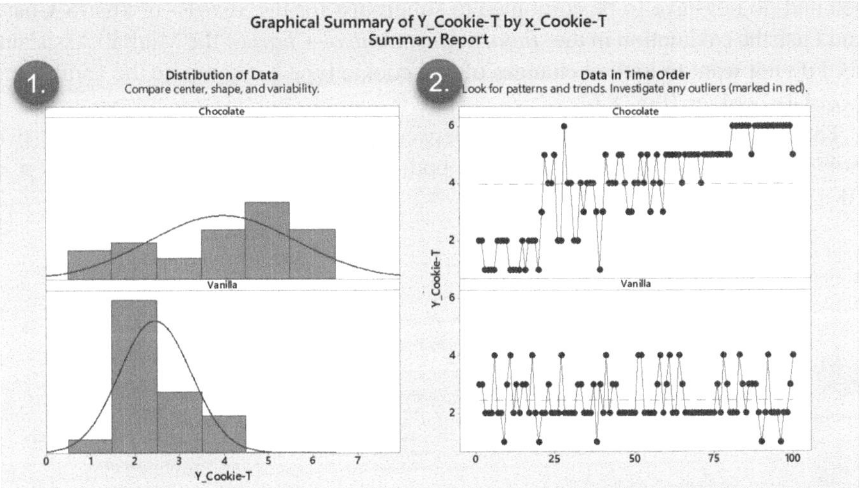

Group	N	Mean	95% CI	StDev	95% CI	Min	Median	Max	Normality Test P	Decision
Chocolate	100	3,98	(3,6536; 4,3064)	1,6452	(1,4445; 1,9112)	1	4	6	<0,005	Fail
Vanilla	100	2,46	(2,3019; 2,6181)	0,79671	(0,6995; 0,9255)	1	2	4	<0,005	Fail

1. Histograms for Chocolate and Vanilla Cookies with the expected overlaid normal distribution curve (N=100), indicating
   a) deviation from the normal distribution (p< 0,005)
   b) different means (3,98 vs. 2,46),
   c) different standard deviations (1,645 vs. 0,796),
   d) within the expected data range from: 1= very good to 6= very bad
2. Time series plots for Chocolate and Vanilla Cookies: the course of data shows, that the rating of Vanilla Cookies is stationary over time, varying around the value of 2, while the rating of the Chocolate-Cookies decreases in the range from 2 (good) to 6 (very bad)

Fig. 8.5 Graphical summary: Distribution and time series of the taste ratings for chocolate and vanilla cookies. The Time-Series-Plot shows that the taste of chocolate cookies (*Y*) decreases over time

and then stacked the data points in the respective categories on the *Y*-axis (see Fig. 8.5).

8.3.1 Process Control

Control-Charts are a special variant of the Time-Series-Plots. I use the I-MR Control-Chart, because the $N = 100$ data points in the chart can still be mapped well and do not have to be combined to subgroups for the xbar/R- or xbar/S-Chart. And I use the calculation in the: *Before/After Control-Chart* of the Minitab Assistant. But I do not want to look at changes of one cookie type but compare the vanilla and chocolate cookies (Fig. 8.6).

There is no significant difference between the standard deviations of the two cookie types ($p = 1$) but the means of both cookie types differ significantly ($p < 0.001$).

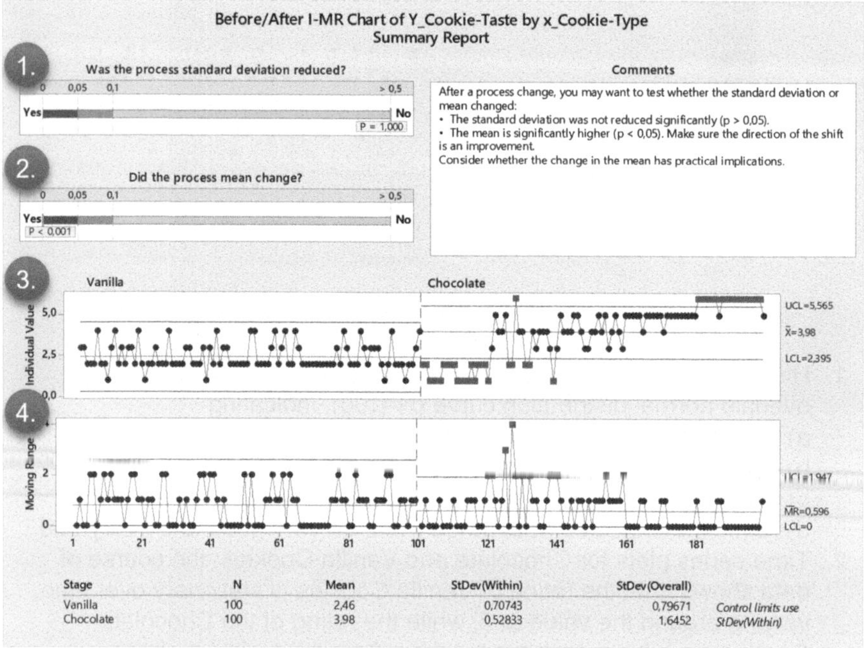

I-MR (**Individual/ Moving Range**) Control Chart for before-after comparisons, applied to compare the Taste of Van.- and Choc.-Cookies

1. The standard deviations of both cookie types do not differ significantly (p= 1)
2. The means of both cookie types differ significantly (p< 0,001)
3. The **Individual Values** chart shows many outliers (red), i.e. ratings that are outside the control limits, overall indicating a shift in the ratings from good to bad and thus a decrease of the performance
4. The **Moving Range** chart shows some noticeable changes from cookie to cookie (red)

Fig. 8.6 I-MR-Chart: The Time-Series-Plot (see also Fig. 8.5) and the I-MR-Chart show that the Taste of chocolate cookies (*Y*) decreases over time

The Individual-Values-Chart shows the time course as in the Time-Series-Plot, plus horizontal lines for the mean and the calculated upper and lower control limits (UCL/LCL).

And it shows many outliers in red, i.e., ratings that are outside the control limits and indicates an overall shift in the ratings from good to bad and thus a decrease of the performance.

It is noticeable that in chocolate cookies the very good rated cookies on the bottom of the chart are also indicated as outliers in red. This is because the control limits were calculated based on the data of the overall poor performing chocolate cookies, which raise the control limits here compared to the vanilla cookies. If I use the Upper Control Limits (UCL) of the controlled process for vanilla cookies and set the Lower Control Limit to zero (LCL = 0), then the bad performance of the chocolate cookies becomes clearer, with even more outliers above the UCL.

Further signals in Control-Charts, such as trends, cycles, shifts, and drifts, are described later in the CONTROL phase. But we can already here make a note of these signals for our Root-Cause-Analysis to see what the causes of these shifts and extreme ratings were. This also applies to the signals in the MR-Chart (Moving-Range-Chart). Here the red data points show strong chronological changes in the ratings of the taste from cookie to cookie. And again, the question of the causes for these fluctuations arises.

8.3.2 Process Capability

Let us next check the process capability of the chocolate cookies. If the data is nominally scaled, e.g., in defect vs. ok, then you can calculate the Yield, Defects per Unit (DPU), Defect Parts per Million (ppm), and the Sigma Level (Fig. 8.7).

I already introduced (Sect. 2.5) the well-known case of 3.4 defects in 1 Mio units, with one opportunity for a defect per unit. This leads to a long-term Sigma Level of 4.5 and a short-term level of 6 sigma.

If you need to calculate the Sigma Level for complex products with several different opportunities for a defect, then the index: Defects per Million Opportunities (DPMO) is appropriate. If you, e.g., assume 100.000 units instead of a million, with 3.4 defects and 10 opportunities for a defect, you will get the same results as calculated before. Every opportunity is treated as a unit and the total number of opportunities result from the number of unit's times the number of opportunities.

If in another case the performance level is given as a percentage value, e.g., Yield, then you can transform it into a Sigma Level and vice versa.

If your data is cardinally scaled, normally distributed and if the specification limits are set, then you can calculate the Pp/Ppk capability indices and the corresponding Sigma Level. In our case, however, the data is only ordinally scaled and not normally distributed. Therefore, the indices may only be interpreted with reservation.

I enter the data of our chocolate cookies here: The lower specification limit: LSL = 1, which also represents one boundary of our rating scale and USL = 4, as

Indices of Process Capability

Calculation of Process Capability based on Units, Defects and Opportunities for Defects			
Definitions	**Symbol**		**Enter Your Data**
Units	U		1.000.000
Defects	D		3,40
Opportunities for a defect	O		1
Number of operation steps	m		
defect Units	**Symbol**	**Calculation**	**Result**
Defects per Unit	DPU	D/ U	0,0000034
Defect Parts per Million	PPM	D/ U x 10^6	3,40
Defects per Unit			
Total Opportunity	TOP	U x O	1.000.000
Defects per Unit Opportunity	DPO	DPU/ O	0,0000034
Defects per million Opportunity	DPMO	DPO x 10^6	3,4000000
without consideration of the Opportunities			
Yield (%)		(1- DPU)* 100	99,9996600
Defect (%)		100 - Yield	0,0003400
for prediction of long-term Sigma-Level from short-term measurement	Sigma-Level (long-term)	z-Value	4,50
for prediction of short-term Sigma-Level from long-term measurement	Sigma-Level (short-term)	z-Value + 1,5	6,00
with consideration of the Opportunities			
Yield (%)		(1- DPO)* 100	99,9996600
Defect (%)		100 - Yield	0,0003400
for prediction of long-term Sigma-Level from short-term measurement	Sigma-Level (long-term)	z-Value	4,50
for prediction of short-term Sigma-Level from long-term measurement	Sigma-Level (short-term)	z-Value + 1,5	6,00

Conversion of Yield% into corresponding z-Values (Sigma-Level) and vice versa			
Enter Your Data		**Sigma-Level (long-term)**	**Yield (%)**
	Conversion: Sigma - Yield	4,50	99,99966023269%
	(Data from long-term study)	0,00	50,00000000000%
		Sigma-Level (short-term)	**Yield (%)**
	Conversion: Process-Sigma - Yield	6,00	99,99966023269%
	(Data from short-term study)	1,50	50,00000000000%

Calculation: Pp/ Ppk and Sigma-Level			
Parameter	**Enter Your Data**	**Sigma-Level (solely based on dispersion - unusual)**	**Pp**
Lower Specification Limit (LSL)	-6,00	6,00	2,00
Upper Specification Limit (USL)	6,00	**Sigma-Level (based on position & dispersion - usual)**	**Ppk**
Mean (xbar)	0,00		
Standard Deviation	1,00	6,00	2,00

Fig. 8.7 sigmaGuide: Calculate different capability indices of your process based on nominal scaled values, percent values, and cardinal scaled values

determined in the Data-Collection-Plan. I enter the calculated mean (= 3.98) and the standard deviation (= 1.65) and get an estimation of Ppk = 0.01 and a corresponding Sigma Level = 0.03.

Both values would even become negative if the mean value were greater than the upper (USL) or below the lower specification limit (LSL). And many projects I have worked on in the past have started with a negative Sigma Level. But it does not matter where you are today, but what you will have achieved tomorrow. Just try this out for your own processes and you will get a first estimation of their capability.

Details Box: Z.bench as an Estimate of Sigma Capability
Statistics programs like Minitab calculate Z.bench values to estimate the Sigma Level. They are based on the normal distribution and convert cumulated percentage values of one side of the normal distribution into Z-values and vice versa (Fig. 8.8).

The results of the *Minitab/Assistant/Capability Analysis* first show again the previous entered upper specification limit (USL = 4) and the target value (= 1.5)

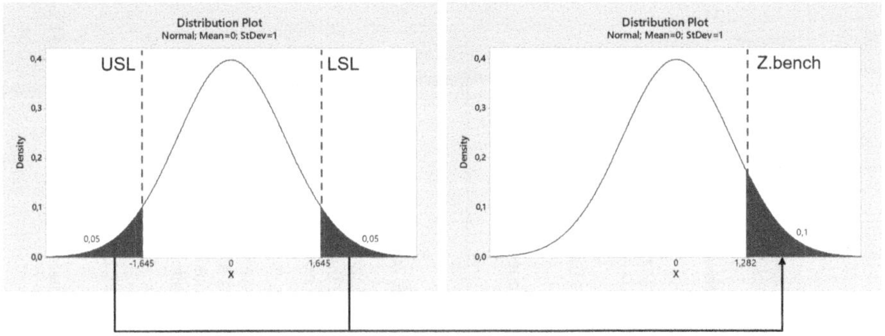

Fig. 8.8 Z.bench as an estimate of the sigma capability level: Cumulated percentage values from both sides of the normal distribution transformed into Z-values and vice versa

from the Data-Collection-Plan. A lower specification limit (LSL) is not necessary in our case, as we only fear ratings greater than 4. And we would be happy if all customers would rate our cookies with a 1, which means: very good and determines the lower limit of our scale.

The actual Sigma Level: Z-actual = 0.01 and the potential Sigma Level—the best value we could expect under the same conditions without changes is: Z-potential= 0.03.

The one-sample *t*-Test shows that the mean (= 3.98) of the ratings significantly differs from the given target value of 1.5 for the taste. This means that we must at least improve the taste of the chocolate cookies (Fig. 8.9).

The Histogram again shows the distribution of our taste ratings. Additionally, one vertical line indicates our target value and another line the upper specification limit (USL). As you can see, the center of the distribution is far from the target value and many cookies are above the upper specification limit (USL). The poor process capability of our chocolate cookies is therefore also graphically traceable here.

Some Belts fear that managers might be upset about the documentation of their process' poor performance. They are therefore considering presenting the results somewhat more positively than they are. But that would not only be dishonest, it also reduces the potential that you can prove in the CONTROL phase.

The following hypothesis tests are aimed precisely at identifying the levers for this potential: the evaluation of the negative influences on the taste of our chocolate cookies.

Questions (SC = Single Choice; MC = Multiple Choice)

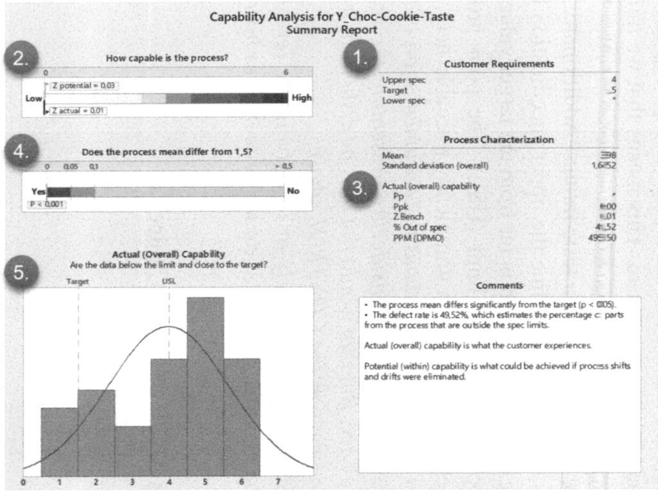

Because the rating data are not normal distributed, the statements about the process capability should be interpreted with caution.

But we are sure, that we have to improve the taste at least of the Chocolate Cookies.

The bimodal distribution with the two peaks indicates, that there were already conditions for a better process capability (see the shift between the first and last cookies in the time series plot).

The process capability analysis shows:

1. The customer requirements are defined as specification limits on the rating scale of taste (see Data Collection Plan)
   - Upper specification limit (USL):= 4
     (Taste should not be rated worse than 4:= adequate)
   - Target:= 1,5 (:= very good/ good)
2. The Z-Values indicate the actual and the potential Process Capability:
   - Z actual= actual Sigma Level= 0,01
   - Z potential= potential Sigma Level= 0,03
   (The rating of taste of Chocolate Cookies is far away from 6 Sigma)
3. Different Parameter of Process Capability:
   - Pp (process performance): not calculated due to the missing LSL
   - Ppk (k= katayori= centre) = 0,00
   - Z.Bench = 0,01
   - % Out of specification = 49,52%
   - PPM / DPMO = 495150
   These parameter focus on different aspects of process capability
4. The (one-sample-t-) Test indicates a significant difference between the mean of the rating of taste (= 3,98) and the target (= 1,5). We can conclude, that our next cookies would also **not** achieve the target, if we do **not** change the process.
5. The histogram shows a bimodal distribution and the Anderson-Darling Normality Test (not shown here) indicates a significant result, meaning that the data for the Chocolate Cookies are also not normally distributed. This means: interpretation of the results under reserve

Fig. 8.9 Statistical tests on the taste of the critical chocolate cookies (Y). Focus is now on the difference of taste between chocolate vs. vanilla cookies (Y)

Video 28/ Ch. 8.3/ #1 (SC)	Please assign the following control-charts to the required scale levels.			
		nominal	ordinal	cardinal
1	u-Chart			
2	xbar-S-Chart			
3	xbar-R-Chart			
4	I-MR-Chart			
5	p-Chart			
5	points			

Video 28/ Ch. 8.3/ #2 (SC)	Please assign the following capability indices to the required scale levels.			
		nominal	ordinal	cardinal
1	Defect-%			
2	Cp/ Cpk			
3	Pp/ Ppk			
4	Defects per Unit			
5	DPMO			
5	points			

Tasks

Business-Projects: Please prepare the necessary Control-Charts and calculate the process capability for your most important *Y*s.

Standard-Project: Please prepare the necessary Control-Charts and calculate the process capability for your most important *Y*s.

Process Capability: The number of trash pieces is cardinally scaled, because 2 bits is twice as many as one bit; therefore, you can use the Cpk/Ppk/Z.bench capability analysis in Mintab. You should not enter a lower specification limit, because 0 is a natural lower limit. As upper limit you can enter a number that seems plausible to you or your Sponsor, e.g., 10. And you can enter: 5 as subgroup size, if you collected 5 observations per cell, as provided in the factorial sampling plan.

Control Charts: You probably have not measured any data over time yet. That is why a control chart based on a Time-Series-Plot does not make sense. Nevertheless you could display your single values in an I-MR-Chart or treat your subsamples with $n = 5$ as subgroups in an xbar-R-Chart.

8.4 Hypotheses Testing[4]

The statistical tests are used to evaluate the significance of the critical Difference and Relationship Hypotheses from the MEASURE phase. In the case of significant results, we will also decide on the practical relevance of their impact strength. Since we identified triggers of the problems in the Input- and Process-Mapping-Analysis, significant and relevant results from the hypotheses tests can serve as starting points for the Root-Cause-Analysis.

[4]Video 29 (https://www.youtube.com/channel/UCShsy50l5dFsoSXU__hBKSA).

The first hypothesis states: There is a or no Difference in the rating of Taste (Y) between different cookie types (vanilla vs. chocolate) (x) (Fig. 8.10).

To distinguish between the cookie types and then focus on the corresponding output with the bigger problem, I chose the 2-sample t-Test, although the data is only ordinally scaled and not normally distributed. The Wilcoxon–Mann–Whitney Test is suitable here without restrictions, but the t-Test is quite robust against violations of these conditions.

We obtain a significant result with a p-value < 0.001, related to the significance level of alpha $= 0.05$, i.e., 5%. The calculated probability of error is therefore smaller than the previously determined typical significance level alpha and therefore we reject the null hypothesis (H0): There is no difference … and accept the alternative hypothesis (HA): There is a difference …

The statistics table shows the size of our sample ($N = 100$), the mean values for chocolate (= 3.98) and vanilla cookies (= 2.46), and the 95% confidence intervals for the means. The true mean value (μ) in the population of all chocolate and vanilla cookies is therefore expected with a 95% probability within the limits of the respective confidence intervals.

These confidence intervals are determined by the accepted alpha (5%) and beta (20%) errors, the sample ($N = 100$) size and the standard deviation ($s = 1.645$ and $s = 0.796$).

The next result shows a difference of: 1.52 between the means of the ratings and the surrounding 95% confidence interval for this difference of: 1.16 to 1.88. We therefore expect—with 95% confidence—that the mean difference of the population lies within this interval. The statistical significance of the test results from the fact that this confidence interval does not include the difference = 0, as postulated in H0. Therefore, we reject the null hypothesis (H0) and accept the alternative hypothesis (HA).

The Interval-Plot graphically illustrates this result. The x-axis shows a vertical line at the difference = 0, the calculated difference between the means for chocolate and vanilla cookies, as well as the surrounding confidence interval. Here it becomes clear that the difference = 0 lies outside the confidence interval, i.e., a difference = 0 is not expected. Instead we expect at least a difference of: 1.16 and in the worst case a difference of: 1.88 between the ratings.

Details: Statistical Significance and Practical Relevance for Differences

Is this statistically significant difference between the cookie types also relevant for our bakery? Is it important for us in practice that chocolate cookies are rated at least 1 rating point lower and on average 1.5 points lower on a six-step rating scale?

Let me explain the question using another example. Imagine if we had only found a difference between the ratings of 0.01 points on the scale. This small

(continued)

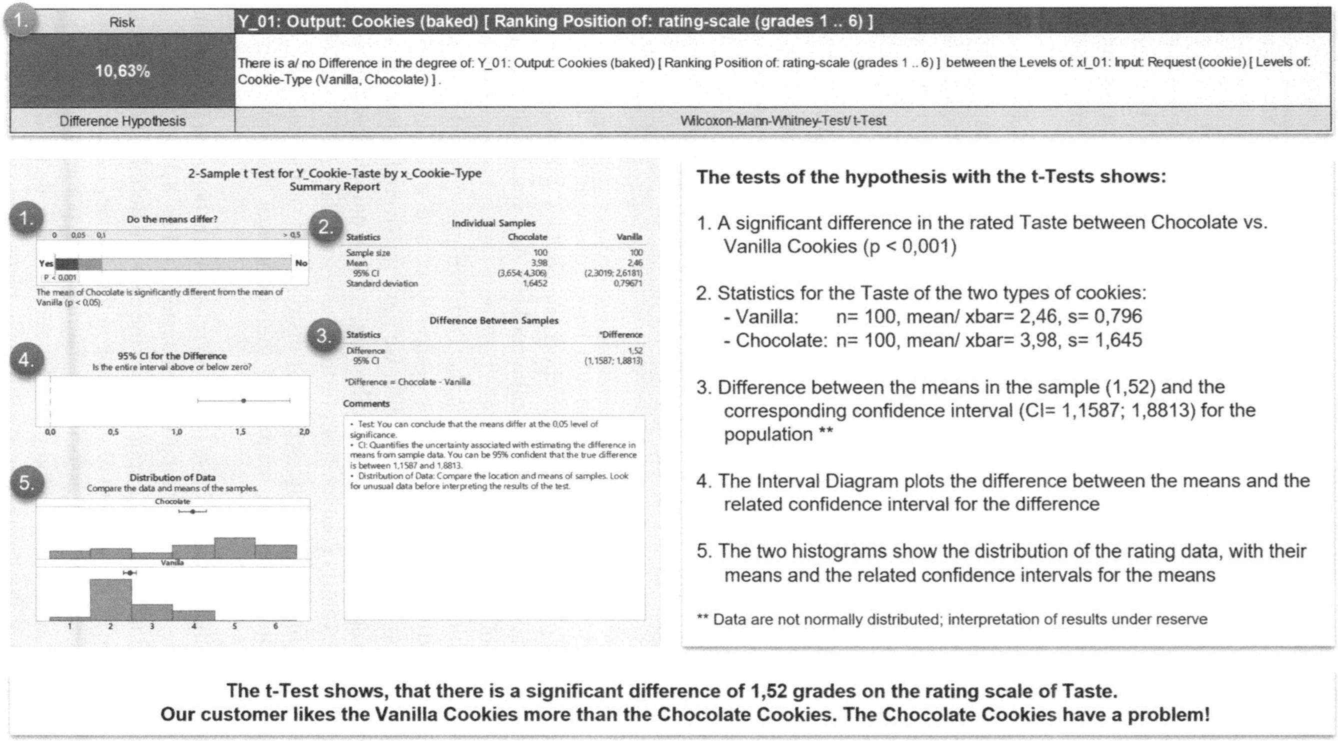

Fig. 8.10 First Hypothesis: There is a difference in taste (Y) between cookie types (x). Focus is now on the negative influences (x) on the taste of chocolate cookies (Y)

difference would not become significant for our sample size of $N = 100$ and we would keep the null hypothesis that there is no difference.

With the statistics tool: *Power and Sample Size* I calculated that with a sample size of $N = 2{,}26{,}000$ this difference would become significant for the given standard deviation and the same alpha- and beta-errors. So, we would have shown a very small difference, this time based on a very large sample, statistically significant and would also reject the H0 and assume the HA. But in practice this small difference probably has no relevance. It is irrelevant in practice that one cookie type is rated on average with: 2 points and another cookie type with: 2.01 points.

Therefore, you should always ask yourself, if you get a statistically significant result, if the difference is relevant for your practice. Does a difference have a practical relevance, even if you base it on the lowest expected difference value of the confidence interval, which is: 1.16 in our case? We will continue this topic in the relationship hypothesis, as well as in the CONTROL phase when it comes to proving a before/after difference.

Here I would like to summarize for the moment that a statistical significance informs us about the resilience of the hypothesis in the future, that in the future we will find a similar difference under similar conditions. And we can only answer the question of the practical relevance if we assess the importance of a certain significant difference or relationship for our further analyses and thus for our company or our customers.

In our case we think that the result should have consequences. If we change nothing, then with a probability of 95% the chocolate cookies will also in future be rated at least 1.15 grades worse than the vanilla cookies. And we can already include this result in our Root-Cause-Analysis.

Our primary problem we had defined in the DEFINE phase is: Y1_cookie taste bad. Now we can differentiate between the cookie types and will focus on the bad taste of chocolate cookies. The differentiation we have uncovered is not yet a cause of our problem but a specification. In the following, we will only deal with chocolate cookies and exclude vanilla cookies from further analyses. They have no problem.

The Histograms finally show the separated distributions of the ratings for both cookie types, with the respective means and the confidence intervals around the means.

The second hypothesis states: There is a or no relationship between the amount of chocolate in a cookie (x) and the rating of their taste (Y) (Fig. 8.11).

To investigate the modeled assumption about the causal determination of the taste of a chocolate cookie by the amount of chocolate in that cookie, I chose the General-Regression to statistically test the associated relationship hypothesis.

The Rank-Correlation would be more suitable for ordinal scaled and not normally distributed ratings, but the Regression-Analysis can better demonstrate the purpose

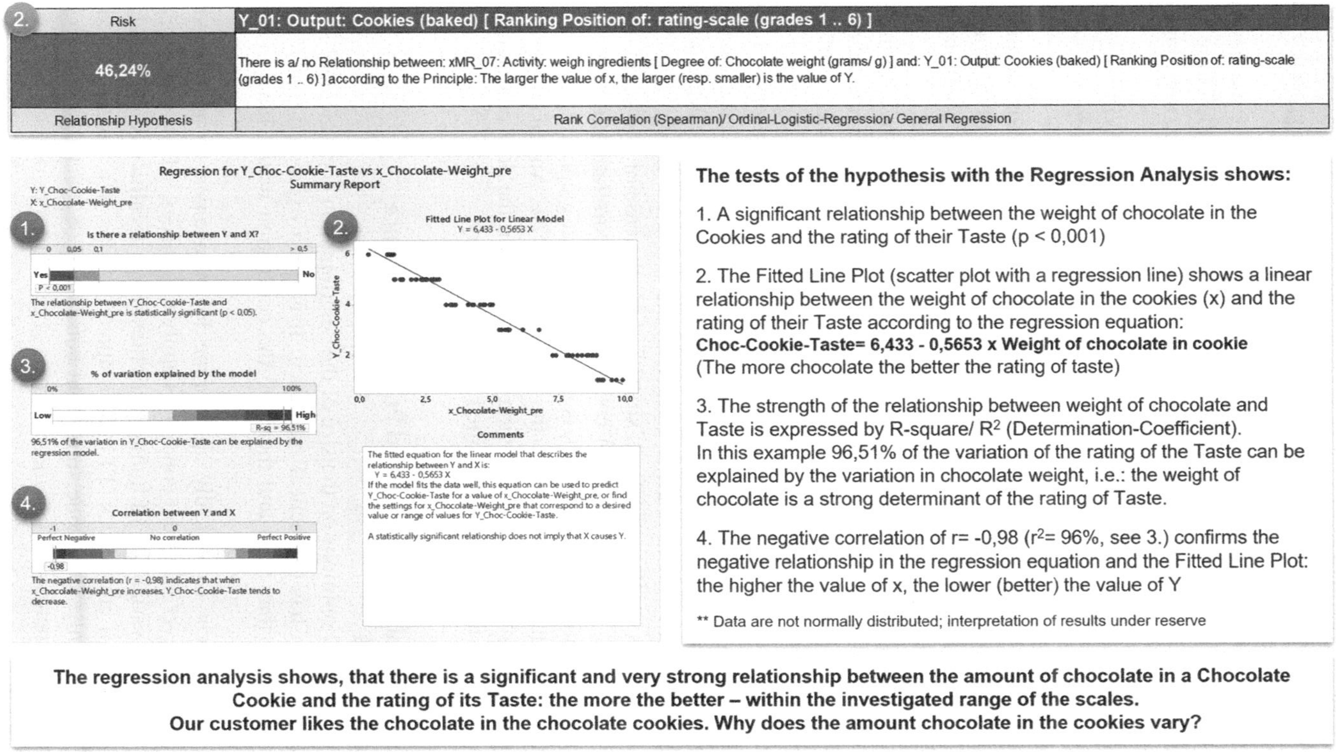

Fig. 8.11 Second Hypothesis: There is a relationship between the amount of chocolate (x) and the rating of Taste (Y). Focus is now on the influences (x) on the variation of the amount of chocolate in chocolate cookies (Y)

of the hypothesis and its test. For this reason, however, the results must-be interpreted with reservations.

First, we get again a significant result with a p-value < 0.001 at a predetermined 5% significance level alpha.

The Fitted-Line-Plot, a Scatter-Plot with a regression line, shows the negative relationship between the amount of chocolate in the cookie (x) and the rating of its taste (Y): The more chocolate in a cookie, the lower the grade value, meaning: the better the rating.

The regression equation expresses this relationship numerically. With this equation: *Y_chocolate cookie-taste = 6.4 − 0.57 * weight-of-chocolate*, we could forecast the taste rating, based on the amount of chocolate in a cookie.

Every regression analysis leads to such a regression equation. But how accurate is such a prediction, how well can we rely on it? This again is a question of the practical relevance. We got a significant result, but only the determination coefficient R-squared (R-sq, R^2) expresses the benefit of the result.

In our case R-sq = 96.51% and thus very high. It means that by varying the amount of chocolate within the selected range, about 97% of the taste variability can be explained. In other words: The taste rating of a cookie is determined to 97% by the amount of chocolate in that cookie.

For our bakery we thus have found a strong, practically relevant cause for the rating of the taste. And of course, I consider this cause in our Root-Cause-Analysis: Why is a chocolate cookie rated bad? Because the amount of chocolate is too low! Or more precise: The lower the amount of chocolate, the worse the taste.

The determination coefficient (R-sq), simplified for the case of one x and one Y, results from squaring of the correlation coefficient. This correlation, in our case a high negative correlation ($r_{xy} = -0.98$), is shown in the last graph. It quantifies the pattern of the fitted line plot and leads to the R-sq value of 97%. And of course, there are also confidence intervals for correlation coefficients and the derived determination coefficients, but these are not specified in this summary.

Details: Statistical Significance and Practical Relevance for Relationships

There are different opinions on the size of the determination coefficient (R-sq) of above which it has a practical relevance. It is often expressed that R-sq should be greater than 0.80 and therefore the correlation would have to be greater than 0.89.

This requirement is basically appropriate for mechanical or automated production. However, different conditions apply to services, whose processes are subject to the variability of human activities. I can recall a case of a high cancellation rate of employees (Y) in a company's shops. We have searched intensively, but only found one significant correlation ($r_{xy} = 0.45$) with the workload due to increased customer volumes at certain times of the year (x).

(continued)

Nevertheless, the resulting determination of R-sq = 20% had a practical relevance for us in the subsequent Root-Cause-Analysis, because we had only very few and very weak explanations for the cancellation rate. So, if we do not know better, then a 20% determination is better than guessing. As with the difference hypotheses, we can only decide subjectively on the practical relevance of a relationship.

The third hypothesis states: There is a or no difference in the taste of a chocolate cookie (Y) between the conditions: The Chef eats the already weighed chocolate from the ingredient vs. not (x). Again, I chose the 2-sample t-Test, although not all conditions for this test are fulfilled here. And we again obtain a significant result with a p-value < 0.001 (Fig. 8.12).

The statistics show a mean value of 4.6, if our chef nibbles chocolate from the ingredients, and if our chef does not nibble on the chocolate, then a mean value of 1.5 results. The difference between the ratings under these two conditions is thus 3.1. The expected true difference in the population (μ) lies with 95% confidence in the interval between: CI $= -3.45$ and -2.75. If our chef nibbles on chocolate from the ingredients, then the cookies will be rated as at least 2.75 points worse and this significant result of course also has a practical relevance for us.

The Interval-Plot illustrates this result again graphically, with the vertical line at the difference = 0, which should apply if the null hypothesis (H0) is valid, and the calculated difference between the two conditions with the surrounding confidence interval.

The Histograms finally show the separated distributions of the ratings under the conditions, with the respective means and the confidence intervals around the means.

The result of this difference hypothesis provides a cause for the different amount of chocolate in the cookies: *Why* does the amount of chocolate vary in the cookies? *Because* our Chef nibbles chocolate from the weighed ingredients. Or vice versa: *If* our chef nibbles chocolate from the ingredients, *then* the amount of chocolate in the cookies is reduced.

Since we have no data about why our Chef nibbles chocolate from the ingredients, we are now again dependent on the opinion of the experts. But of course, we already have an assumption why that is. Our next task is to combine the statistical results with considerations of the experts in a Root-Cause-Analysis.

Questions (SC = Single Choice; MC = Multiple Choice)

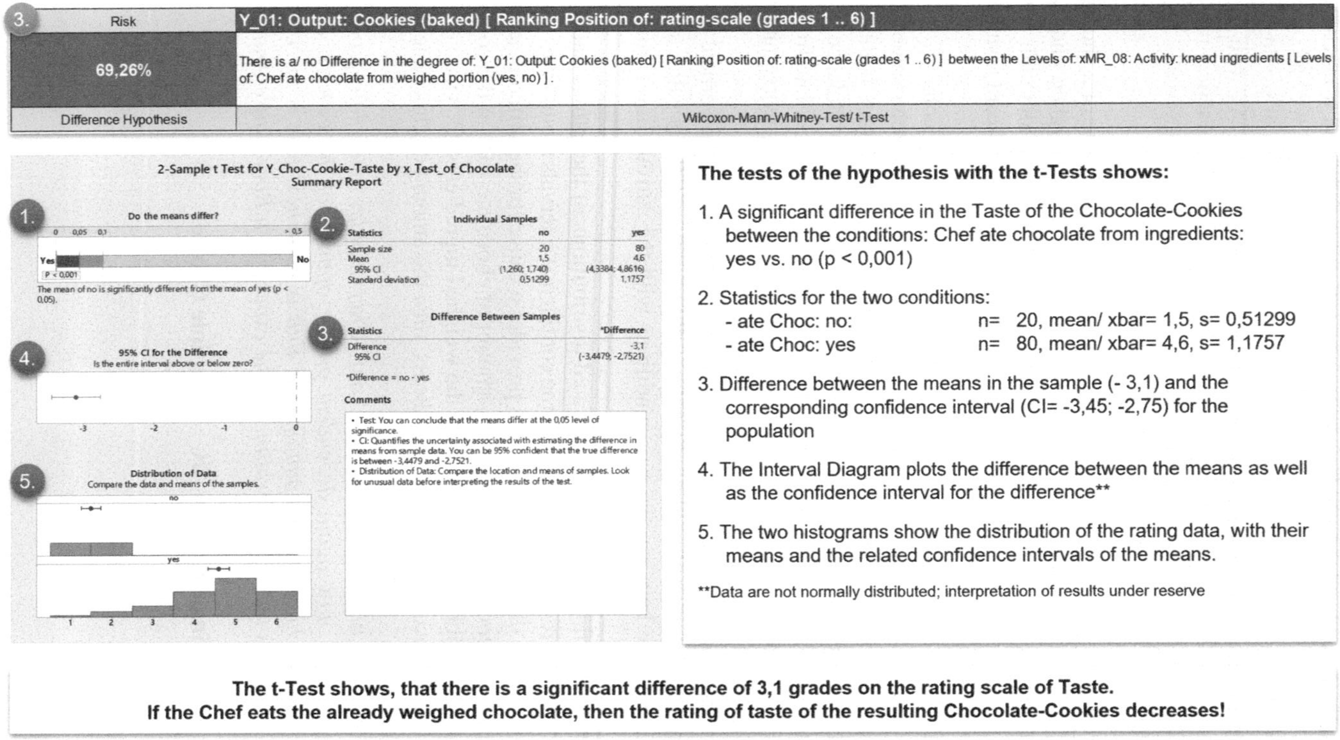

Fig. 8.12 Third Hypothesis: There is a difference in the taste of a chocolate cookie (Y) between the conditions the Chef eats the chocolate from ingredient vs. not (x). Focus is now on the root causes (x') of eating chocolate from the ingredients (x)

Video 29/ Ch. 8.4/ #1 (MC)	Please assign the required scale levels for x and Y to the statistical tests.					
		x: nominal/ 2 levels	x: nominal/ > 2 levels	x: cardinal	Y: nominal	Y: cardinal
1	t-test					
2	Chi-Squared Test					
3	Correlation					
4	ANOVA					
5	Regression-Analysis					
5	points					

Video 29/ Ch. 8.4/ #2 (SC)	Which alpha levels are suitable for which type of examination?			
		alpha= 10%	alpha= 5%	alpha= 1%
1	Preliminary investigation of an assumption			
2	Standard decision			
3	Major decision			
3	points			

Video 29/ Ch. 8.4/ #3 (SC)	Which of these statistical test data determines the ...					
		p-value	alpha	difference between means	level of correlation	level of determination
1	statistical significance					
2	practical relevance					
2	points					

Tasks

Business-Projects: Please calculate the statistical tests for your most important hypotheses. Perhaps now that the data and graphical evaluations are available, you are developing more interesting hypotheses and testing them statistically. In practice, this inductive approach very often complements the deductive derivation of hypotheses from the C&E matrix.

Standard-Project: Please calculate the statistical tests for your most important hypotheses. Perhaps now that the data and graphical evaluations are available, you are developing more interesting hypotheses and testing them statistically. In practice, this inductive approach very often complements the deductive derivation of hypotheses from the C&E matrix.

8.5 Root-Cause-Analysis

8.5.1 Preparation[5]

In our bakery there are only a few problems and also the number of influences is small. Nevertheless, we have not been able to measure data for all hypotheses, because this would have been very time consuming. In our project, we could now immediately get into the Root-Cause-Analysis and link statistical results with expert assumptions. Nevertheless, I would like to prioritize the problems first to better organize the further procedure.

We are committed to solving all the problems of the Project-Charter and to achieving the related targets. And a separate Root-Cause-Analysis must be created for each problem. Especially in complex projects it can be efficient to prioritize the problems for their analysis.

The basic prioritization criterion is the severity of the problems we identified in the DEFINE phase (Voice-to-Critical). Now I would like to additionally distinguish between problems of different categories (quality, availability, consumption) and problems within the same category.

According to a well-known lean principle, quality problems have the highest priority because it makes no sense to make an error faster. Typically, a quality problem leads to a reduction in availability. If an order is incomplete, we need to clarify it with the customer. And this clarification increases the cycle time and the resource consumption. Solving the quality problem would thus increase the availability and decrease the consumption of human resources. The same applies to all other error corrections. These problems are therefore interdependent (Fig. 8.13).

However, solving the quality problem alone does not lead to a complete solution of the availability or resource consumption problem. Availability problems often result from the sum of many small delays in several activities, while quality problems are often triggered by exactly one activity.

Relationships between problems can be well discovered with plausibility considerations and confirmed with correlation or regression analysis. This also applies to dependencies within a category.

The Pareto-Chart is suitable to prioritize the problems based on their frequency or based on the resulting quality costs. I collected data to the quality problems: Cookie taste bad, cookie weight wrong, cookies diameter wrong, ingredients weight wrong, cookies broken, cookies burnt, and ingredients expiry date exceeded.

The Pareto-Chart is suitable for nominal scaled data, e.g., ok vs. defect. To transform the ordinally scaled rating data to the lower nominal scale level, I assigned the ratings above our specification limit 4 to the category: ok and the bad ratings 5 and 6 to the category: defect (Fig. 8.14).

Of course, this also works for cardinally scaled data. This way you can display and prioritize all problems in a Pareto-Chart.

[5]Video 30 (https://www.youtube.com/channel/UCShsy50l5dFsoSXU__hBKSA).

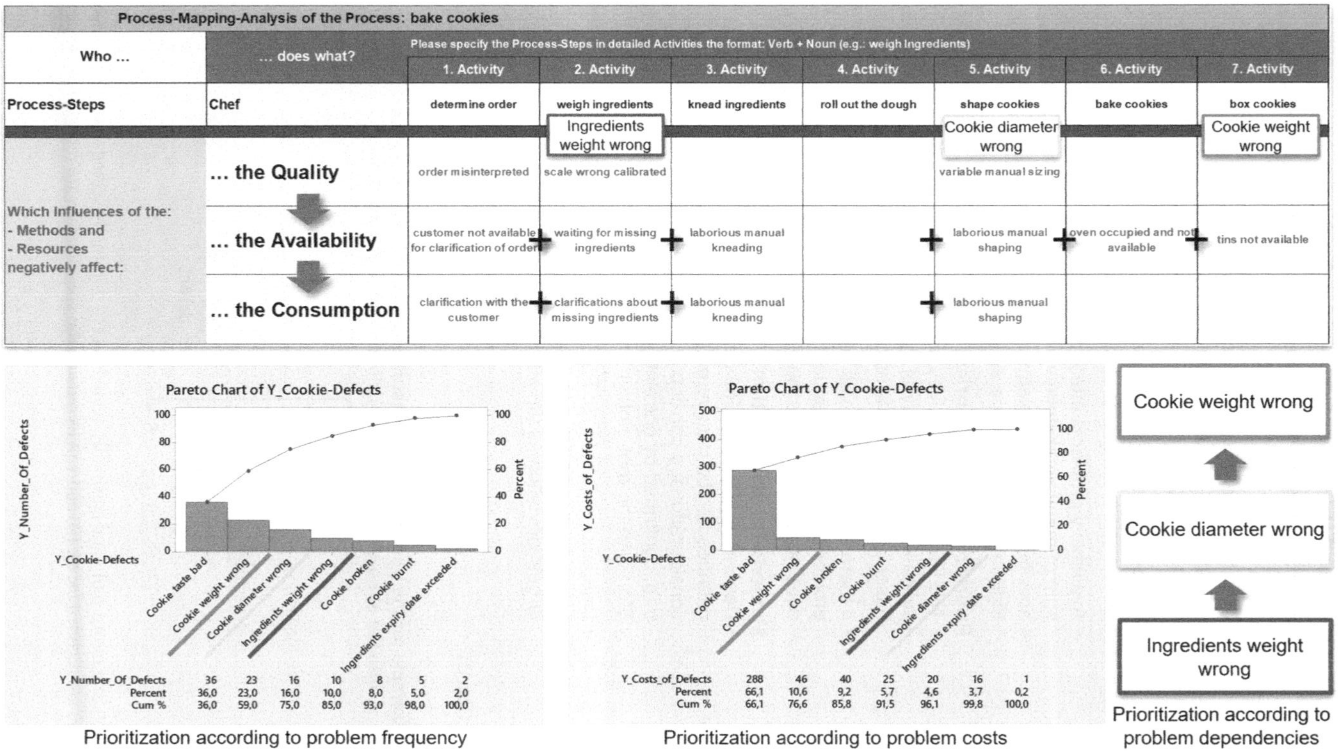

Fig. 8.13 Analysis strategy: Prioritize the problems and focus on important problems between and within problem categories (quality, availability, consumption)

Basic prioritization criterion:
- Severity of Problems (Voice-to-Critical)

Prioritize between categories:
- Quality problems have priority
- Quality problems can lead to availability problems for necessary corrections and feedback loops
- Corrections and feedback loops consume at least human resources
- Availability problems often result from the sum of many small delays in several activities. Quality problems are often triggered by exactly one activity

Prioritize within categories:
- The Pareto diagram gives an initial indication of the priority of the problems based on their frequencies or costs
- If there are dependencies between the problems, the most frequent or expensive does not need to have the highest priority

Methods for prioritization:
- Depended problems: Relationships and priorities can be uncovered by plausibility considerations, a corresponding hierarchy tree, confirmed by correlation/ regression. Start with the root problem.
- Independent problems: Pareto chart, Chi^2-test, t-Test or ANOVA help to identify the largest deviation as the starting point.

Fig. 8.13 (continued)

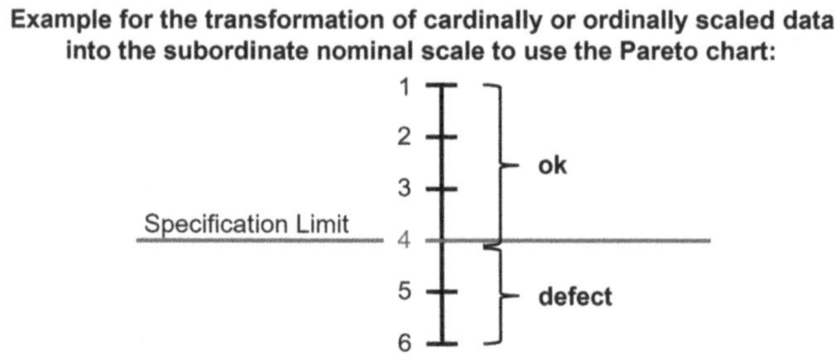

Fig. 8.14 Recoding of ordinally and cardinally scaled data to nominally scaled data: Values within the specification are recoded to: ok (0) and values outside the specification are recoded to: defect (1)

The left Pareto-Chart prioritizes the defects according to their frequency and the right according to the resulting quality costs (see Fig. 8.13). Our problem with the highest severity: Cookie taste bad, is in first place in both charts. But there are a few shifts in the lower rankings. And you can choose the criterion, frequency, or cost, according to which you work through the ranking. However, this only applies if the problems are independent.

In the case of dependent problems, the most frequent problem often occurs at the end of the process chain. If, e.g., the amount of ingredients is too small, then the probability increases that the cookies will be shaped smaller, to keep the total number constant and that the required weight of the cookies will then not be correct.

In addition, there may be further influences leading to the cookies being too small and their weight deviating. The problem of: Ingredient amount too small, then receives a subordinate place in the Pareto-Chart but should be solved with higher priority. The solution of the ingredients problem will probably reduce the problem of the cookie diameter and the problem of the cookie weight.

To uncover these dependencies, you can first create Hierarchy-Tree, just for showing the dependencies between problems, based on observations and logical assumptions and if possible, confirmed by means of correlation or regression. And of course, you can additionally prioritize problems by differentiating them with the *t*-Test, as we did it with the ratings for vanilla vs. chocolate cookies and with the ANOVA, to compare more than two products or services.

Once you have prioritized the problems, you can prepare the result, e.g., the Pareto-Charts and the Hierarchy-Tree on a flipchart for your Root-Cause-Analysis workshop, to discuss it with the experts.

Questions (SC = Single Choice; MC = Multiple Choice)

Video 30/ Ch. 8.5.1/ #1 (MC)	Which statement apply to the prioritization of problems?	
1	The Allows Root-Cause-Analysis to begin with the most important problem	
2	Dependent problems may not need to be solved	
3	The root problems often occur less frequently than the common problems	
4	Prioritization can be based on frequencies and logical dependencies	
5	Logical and frequency-based prioritization leads to identical results	
6	In the case of dependent problems, the Root-Cause-Analysis may be omitted	
7	Root-Cause-Analysis should always start with the problem with the highest non-conformity	
7	points	

Tasks

Business-Projects: Please prepare the Root-Cause-Analysis. Create Pareto-Charts and, if possible, arrange the defined problems in a hierarchical order.

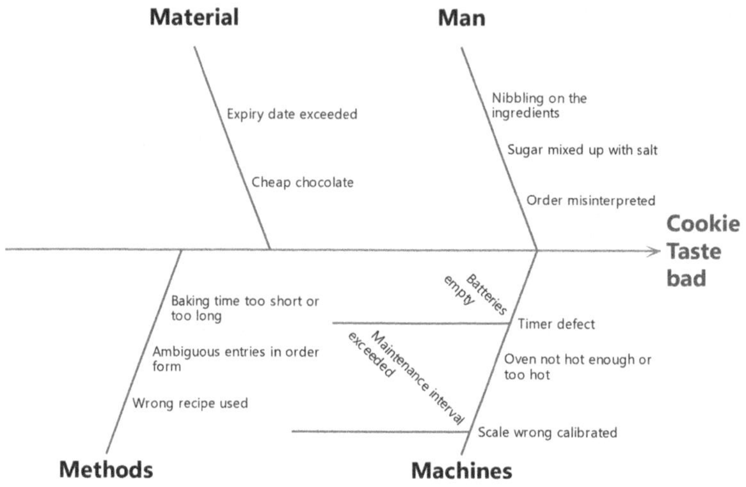

Fig. 8.15 Ishikawa-/Fishbone-Diagram: Category-based (Material, . . .) cause search (in contrast to the chronological search for causes in the Hierarchy-Tree)

Standard-Project: Please prepare the Root-Cause-Analysis. Create Pareto-Charts and arrange at least the problems: (a) garbage removal incomplete, (b) garbage removal effort high, and (c) number of garbage pieces around trash cans >10 in a hierarchical order.

8.5.2 Implementation[6]

The Root-Cause-Analysis continues the hypothesis tests, only without data because any assumption by experts about causes is a hypothesis. In the best case we can identify the root causes based only on statistically tested hypotheses, as we did it already for the important triggers. Unfortunately, data are rarely available for the causes of triggers and their causes and could only be measured with a great effort. Nevertheless, in many cases we can rely on expert observations and their assumptions about causes of the causes.

You probably already know the Ishikawa-/Fishbone-Diagram as a tool for the Root-Cause-Analysis. However, its associative procedure has weaknesses to derive the root causes of a problem chronologically. The Ishikawa-Diagram does not start with the search for the triggers of the problem. Instead, all relevant areas of influence: man (= resource), machine (= resource), material (= input) and method, are first searched consecutively for possible causes (Fig. 8.15).

[6]Video 31 (https://www.youtube.com/channel/UCShsy50l5dFsoSXU__hBKSA).

This type of search with the Ishikawa-Diagram ensures that no possible area of influence is overlooked. As a result, however, trivial associations are often mentioned, which in reality only have a small or no influence at all on the problem. And then the identified root causes, suspected some *bones* deeper in the *fishbone*, will also have little to do with reality. Furthermore, deep branching (by 5 Why's) and necessary cross connections make an Ishikawa-Diagram quickly unclear.

In contrast, Hierarchy-Trees are very well suited to structure the occurrence of a problem, step-by-step and comprehensibly, from its trigger to its root causes.

And I want to additionally integrate our statistical results in the cascade of: *Why ...?—Because ...!* structure and thus link quantitative evidence with rational assumptions into a common analysis strategy to build the Hierarchy-Tree (Fig. 8.16).

In the workshop I explain the experts of our bakery that I started with the primary problem: Cookie taste bad. With the *t*-Test I found out that primarily chocolate cookies are affected and illustrate this focus by dividing the overall problem. With the regression analysis I then show, for the first cause-level, that the taste of chocolate cookies depends on the amount of chocolate in these cookies. To explain why the amount of chocolate varies between the cookies, I refer to the result of another *t*-Test, showing that the taste ratings depend on whether our Chef nibbles chocolate on the ingredients or not.

One expert doubts this conclusion because the *t*-Test is not related to the difference in the amount of chocolate between the conditions. It could also be that the Chef always only nibbles very small quantities, but that a more important cause might be that the kitchen scale is wrongly calibrated.

This is a legitimate objection. And the expert had already mentioned the negative influence of the scale in the Process-Analysis. For me, this results in two consequences:

1. I spontaneously formulate the hypothesis: There is a difference in: the amount of chocolate in the cookies (Y) between the conditions: Chef nibbles chocolate vs. not (x). The evaluation of the existing data with a *t*-Test directly in the workshop again shows a significant result. And with 5.5 rating points difference between the conditions, the result is also practically relevant: The Chef's nibbling has a significant and relevant effect on the amount of chocolate in the cookies. This confirms my previous analysis (Fig. 8.17).
2. And on the same level I note the additional cause: Scale wrong balanced. We cannot rule this out here and now, because we have no data. Thus, I will not break down this cause further at first, because it is pointless to ask why the scale is not calibrated. And I suggest a reality check with a measurement system analysis (MSA).

I proceed similarly with the presumed cause of the exceeded expiry dates of the ingredients. First, data should also here be collected.

On the next level, I ask *why* our Chef nibbles the chocolate from the ingredients. We have not yet formulated a hypothesis for this assumption, nor have we collected any data. And it would be a lot of work for us to collect the necessary data. In

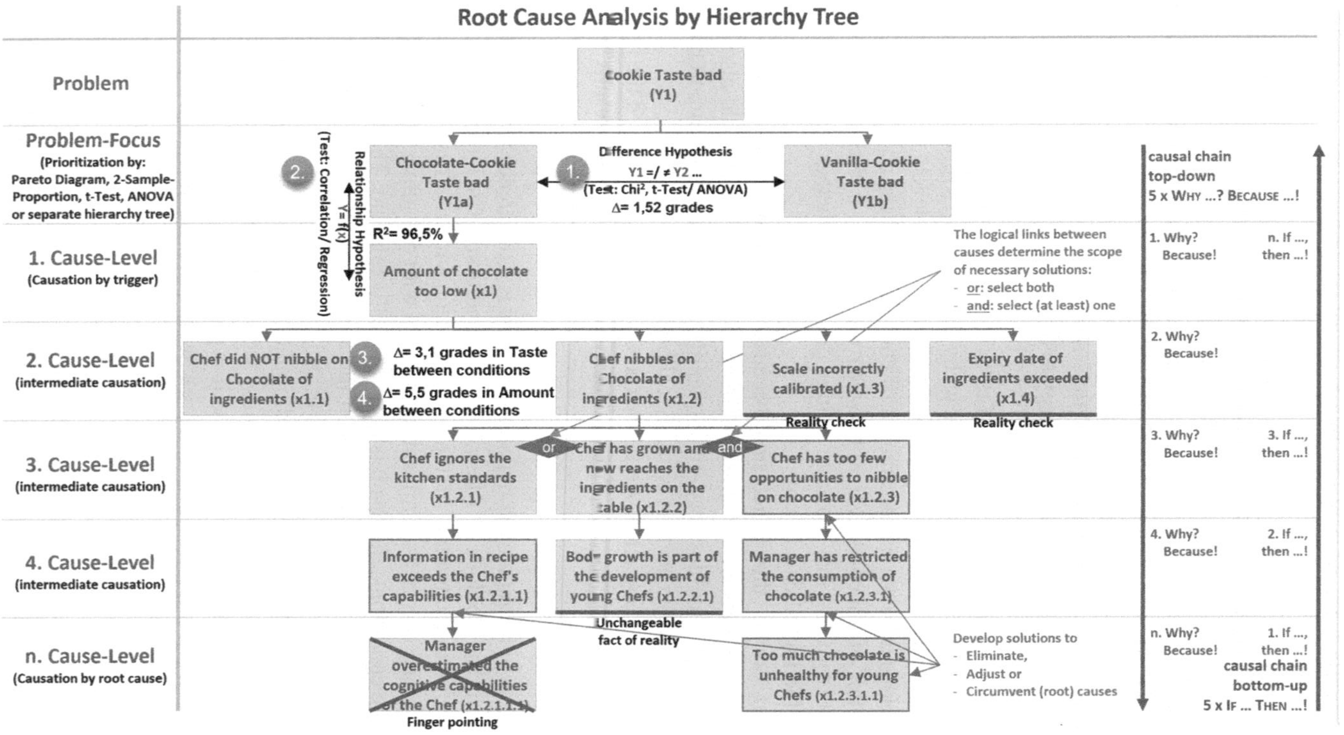

Fig. 8.16 Analysis strategy: Combine quantitative statistical results and logical arguments to identify root causes by (statistically confirmed) assumptions about causal relationships in a Hierarchy-Tree

Analysis strategy

- Create a tree for each problem
- Identify the trigger for this problem and break it down to its root causes
- Combine statistical results from the hypotheses tests with experts assumptions, if no data are available
- Ask "Why?" questions to a given cause and note the "Because!" answers as assumptions of underlying causes
- Typically, 5 "Why -Because" levels are recommended to identify the root causes
- Test the assumptions statistically if corresponding data are available
- For quality problems sometimes more than 5 cause-levels are needed
- Availability and consumption problems sometimes need less than 3 levels
- Stop the analysis (of a branch) if:
  - Reality checks are necessary
  - Root causes are identified
  - The next cause will be related to a person which allows finger pointing
  - unchangeable facts of reality are identified
  - Identified causes explain 80% of the problem frequency
- Question assumed causes that contain: no, not, none or the prefix un- (**Solutions?**)

Fig. 8.16 (continued)

practice, it very often happens that no data is available for subordinate levels of the problem–cause tree. Therefore, we must now rely on observations and rational analyses from experts to identify the root causes.

And our experts name three causes for the Chef's nibbling: Chef ignores the kitchen standards (x1.2.1), the Chef has grown and now reaches the ingredients on the table (x1.2.3) and the Chef has too few opportunities, to nibble on chocolate (x1.2.3).

The cause for ignoring the recipe (x1.2.1) is presumably that the: Information in the recipe exceeds the Chef's capabilities (x1.2.1.1). And it is pointless to ask *why* this is. And it can lead to finger pointing, if we confess, that the manager overestimated the cognitive capabilities of the Chef. (x1.2.3.1.1).

4. There is a difference in: (Y)	between: Levels of (x)	Scale Level of Y	Scale Level of x	Graphical Representation:	Statistical Test:
The amount of chocolate in (x) in the Chocolate Cookies	*Chef eats chocolate from the ingredients vs. not (x)*	*cardinal*	*nominal*	*Box-Plot*	*t-Test*

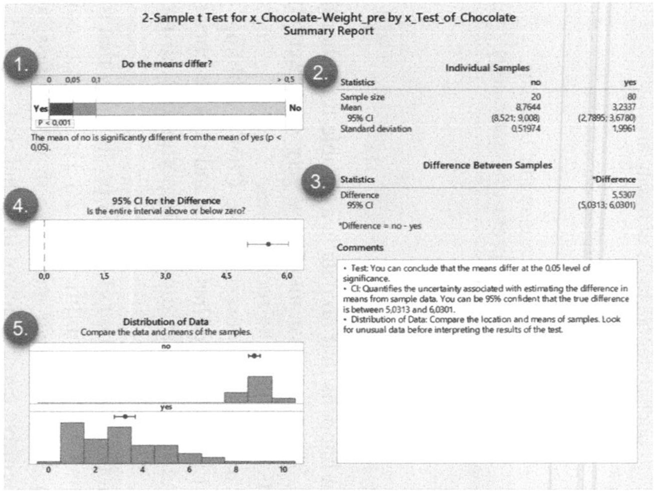

The tests of the hypothesis with the t-Tests shows:

1. A significant difference in: the amount of chocolate in the Chocolate Cookies (x) between the conditions: Chef eats chocolate from the ingredients: yes vs. no (x) (p < 0,001)

2. Statistics for the two test conditions:
   - ate Choc: no: n= 20, mean/ xbar= 8,76; s= 0,51
   - ate Choc: yes n= 80, mean/ xbar= 3,23; s= 1,99

3. Difference between the means in the sample (5,53) and the corresponding confidence interval (CI= 5,03; 6,03) for the population

4. The Interval Diagram plots the difference between the means as well as the confidence interval for the difference**

5. The two histograms show the distributions of the rating data, with their means and the related confidence intervals of the means.

**Data are not normally distributed; interpretation of results under reserve

The t-Test shows, that there is a significant difference of 5,5 grams chocolate in Chocolates Cookies.
If the Chef eats chocolate from the ingredients, then the weight of chocolate in Chocolate Cookies decreases!

Fig. 8.17 Fourth Hypothesis: There is a difference in: The amount of chocolate in (x) in the chocolate cookies between the conditions: Chef eats the chocolate from the ingredients vs. not (x). The result confirms: Eating chocolate from the ingredients (x) reduces the taste of cookies (Y)

Please avoid this finger pointing. Stop the analysis before you get to that level. And even without pointing fingers, we already know that we must adapt the standards for our little Chef. And that is where we want to go: To find causes that we can eliminate, adjust, or circumvent with solutions.

As next cause for the nibbling of the chocolate, it is assumed that the Chef has grown and can now reach the ingredients on the table (x1.2.3). This assumption is also plausible, because it explains the negative trend in the rating of cookies, as shown in the Time-Series-Plot (see Figs. 8.5 and 8.6). The following question on the next level, why the Chef has grown, is again pointless, because body growth is an unchangeable fact of reality (x1.2.2.1). We cannot and will not change that.

However, we know that our Chef has few options to nibble on chocolate (x1.2.3) because her manager has restricted the consumption of chocolate (x1.2.3.1). And the cause for this is that too much chocolate is unhealthy for little Chefs (x1.2.3.1.1).

There is a rule of thumb that the search for root cause search can be terminated, if the elimination of the identified causes would reduce the problem frequency by more than 80%. Therefore, I ask the experts: "How often, in percent, will the problem probably still occur, if the causes of this analysis were eliminated?" The experts cannot think of any further relevant causes. And also in my view, we have sufficiently resolved the problem into its root causes, and I finish the analysis here.

Details: 5 × *Why?* – Necessary and Sufficient to Detect the Root Causes?

The standard recommendation to ask five *why* questions in a cascade to advance to the root causes is basically sensible for the associated solutions to be effective. However, I would not make them mandatory, because there are:

1. Causes that should **first be checked in reality**, before they are assumed as true, e.g., the wrong calibrated scale.
2. Causes whose **underlying causes are trivial**, e.g., kids ignore recipes.
3. Causes that **uncover** an employee's **guilt** with finger pointing. This is the case, for example, if employees develop their own critical standards for carrying out activities based on their preferences. Of course, it should always be asked whether an official standard exists, what the advantages are of deviating from this standard and what might make the application of the standard more difficult.
4. Causes which represent **unchangeable facts of reality**, for which there are no suitable solutions to eliminate them, e.g., kids grow.

Basically, I have made the experience that causes of quality problems are more deeply nested than availability problems. I have often observed that a quality problem resulted from exactly one trigger. And the root causes of the

(continued)

trigger were hidden several levels deeper in the cause hierarchy. In contrast, an availability problem often results from time losses at many activities of the process, whereby the chain from the trigger to the root causes is rather short and transparent. Sometimes only three hierarchy levels in trees for availability problems are necessary.

It is ambitious, but sometimes useful, to additionally detect the logical connections between causes. The causes: Chef ignores the kitchen standards (x_1.2.1); Chef has grown and now reaches the ingredients on the table (x1.2.2) are linked with a logical: **OR**. This means that both causes can occur alternatively and in parallel to explain the loss of chocolate. If causes are linked with a logical: OR, then both have to be eliminated to eliminate the problem.

If on the other hand two causes are linked with a logical: **AND**, then only one of the roots needs to be eliminated. For example, the causes: Chef has grown, now reaches the ingredients on the table (x1.2.2); Chef has too few opportunities to nibble on chocolate (x1.2.3) are linked with a logical AND.

To test the plausibility of our causal chain, we now go back the way, from bottom to top, with a sequence of: *If*..., *then* ... conclusions. It is a reversal of the *Why ...? -Because ...!* procedure. So, I ask: "*If* too much chocolate is unhealthy, is it *then* plausible that, the manager forbids the Chef, to eat too much chocolate?" If the experts find the conclusion plausible, then I continue: "*If* the manager forbids the Chef to eat too much chocolate, does the Chef *then* has too few opportunities to nibble chocolate?" And finally: "*If* the Chef has too few opportunities to nibble chocolate, is it plausible, that the Chef *then* nibbles chocolate from the ingredients?" We can continue this review of our analysis up to the problem and repeat it for each branch of the Hierarchy-Tree.

This check can help to identify gaps in the chain of causes. Please be very vigilant as to whether the conclusions from level to level are direct and valid, so that one argument directly leads to the other and that you can really understand the assumption. If not, then further resolve the assumption by inserting an intermediate cause level or correct the misleading assumption.

Details: Non-existent Solutions Are Not Causes

The review can also help to identify implausible arguments. Let me give you an example for this. I had the problem of repeatedly appearing late at work (problem). My manager asks me: "*Why* do you arrive late"? And I answered: "*Because* I don't have a green Ferrari" (cause). My manager then immediately realized that was very sorry and bought me a red Ferrari (solution). And afterwards I was always on time with my new car.

(continued)

Which statements in this story are false?—Only with green Ferraris I would arrive on time? Or my manager did not buy me a Ferrari at all?—Sure, and he justified it with the fact that the reversal of the statement is wrong: "*If* you don't have a Ferrari, *then* you'll arrive too late."

Most of us arrive on time without a Ferrari. The implausibility can be explained with the Process-Problem-Solving-Model (Sects. 2.4.2 and 2.4.3). Which category does the: Ferrari belong to? Is it a problem, an effect, a cause or a solution?

Another example: "*Why* were the cookies delivered too early?"—"*Because* our chef doesn't use a software calendar."—What is wrong with this conclusion? Which category does the: Software calendar in the model belong to?

The Ferrari and the software calendar are both solutions that do not yet exist. The assumptions express that a problem occurs because a solution has not yet been implemented. And that is nonsense. Instead, we look for causes that really occurred in the past and can be observed in reality. You can see that I am riding a BMW or my bike. But you cannot see that I do not drive a Ferrari.

I have experienced many workshops with this kind of misleading argumentation. People often do not describe what is the case, but what is not yet the case. And I can understand that because it is difficult to observe the world and describe it as it is. Especially if a solution already seems plausible, then the absence of this solution is often presented as a cause. And then a red Ferrari was bought, but they were still late for their meetings.

Perhaps the purchase of the expensive software calendar was not necessary either, because it might have been enough to tell our Chef to tear off only one sheet of the paper calendar every day.

Root causes are the interfaces for solutions. Wrong causes lead to wrong solutions. And the two examples show how important the identification of real causes is for the development of solutions. So please note: If an assumption is made about a cause, that contains: *no*, *not*, *none* or the prefix: *un*, then it is likely that an implicit proposal for a solution is made.

When you have identified all the causes of a problem, mark them. Now you have fulfilled the purpose of the ANALYSE phase for the first problem. You must now repeat this analysis for the other defined problems. And that can be very laborious. But maybe prioritizing the problems, as described in the preparation (Sect. 8.5.1), and the resulting cross-connections between the Hierarchy-Trees will help you save time.

The task in the following IMPROVE phase will be to eliminate, adjust, or circumvent these root causes. And maybe you can win the experts in the workshop to additionally develop solutions right now. This involvement will assure acceptance and support for their implementation.

Questions (SC = Single Choice; MC = Multiple Choice)

Video 31/ Ch. 8.5.2/ #1 (SC) **Which attributes do the Fisbone/Ishikawa diagram and the hierarchy tree have?**

		Fishbone Diagram	Hierarchy Tree
1	Collect information on all influence areas		
2	Identifies the trigger		
3	Collects sometimes irrelevant information		
4	Uses the associative capabilities of experts		
5	Forces to detect chronological dependencies		
6	Allows to integrate statistical tests		
6	points		

Video 31/ Ch. 8.5.2/ #2 (SC) **What logical relationships do these statistical tests represent?**

		If ... then ...	The ... the ...
1	Chi-Squared Test		
2	Correlation		
3	Regression-Analysis		
3	points		

Video 31/ Ch. 8.5.2/ #3 (SC) **What does the word: not or the prefix: un- often indicate in a Root-Cause-Analysis?**

1	Root-Cause	
2	Trigger	
3	Problem	
4	Effect	
5	Solution	
5	points	

Tasks

Business-Projects: Please conduct Root-Cause-Analyses for all problems, at least for the problem on the lowest level of the problem hierarchy. Make sure that the root causes are plausible and so concrete that solutions are already apparent.

Copy the results of all ANLYSE tools and the pictures from the workshop in your Project-Story-Book and comment on the results.

Standard-Project: Please conduct Root-Cause-Analyses for all problems, at least for the problem on the lowest level of the problem hierarchy. Make sure that the root causes are plausible and so concrete that solutions are already apparent. Maybe you can conduct a small workshop with the experts of your local garbage removal service. And maybe they are even ready to develop some solutions together with you (Sect. 9.2.2).

Copy the results of all ANLYSE tools and, if available, the pictures from the workshop in your Project-Story-Book and comment on the results.

IMPROVE 9

9.1 Overview: Steps, Tools, Their Purpose, and Results[1]

With the ANALYSE phase you have completed the most sophisticated part of your project. But you have not improved anything yet. This will happen now in the IMPROVE phase. And the tasks here are less intellectually demanding. Nevertheless, it can be a challenge for you and the expert team to develop suitable and accepted solutions and to implement the derived measures together (Fig. 9.1).

First, the selected causes of the Root-Cause-Analysis are focused, and solution ideas are developed to eliminate, adjust, or to circumvent them.

Discuss the advantages and disadvantages of the solution ideas and, as far as possible, combine the ideas into beneficial, practicable solutions and evaluate them according to their effort/benefit ratio.

Then it becomes concrete and sometimes tough: Specify concrete measures for the selected solutions and define: *Who? Does what? Until when?* Now identify the risks that may arise from the measures and look for suitable countermeasures.

Finally, the Sponsor and the process owners decide which measures should be implemented. The Sponsor and the process owners should also make a clear statement that they want to actively support and monitor their implementation.

Now you are ready to go. Resistance, whether from experts, employees, or managers, should be taken seriously throughout the whole phase. Resistance is often expressed in killer phrases, like: "I don't have time for this!" or "That never works!" Include all affected stakeholders in your communication plan and make sure that the support is always notably stronger than the resistance.

[1]Video 32 (https://www.youtube.com/channel/UCShsy50l5dFsoSXU__hBKSA)

© Springer Nature Switzerland AG 2019

R. Hutwelker, *Six Sigma Green Belt Certification Project*, Management for Professionals, https://doi.org/10.1007/978-3-030-31915-1_9

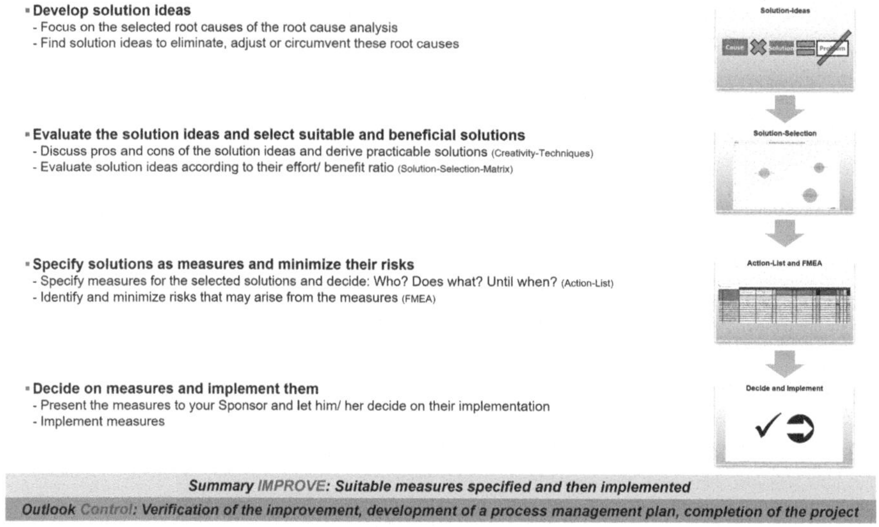

Fig. 9.1 Overview IMPROVE: From the development of solution ideas to the implementation of measures

9.2 Development and Selection of Solutions[2]

The root causes, sometimes together with causes at higher levels, provide interfaces for solutions. If the identified chain of causes actually triggers the defined problem, then a solution should eliminate, adjust, or circumvent the root causes, so that the problem frequency decreases (Fig. 9.2)*:*

- *If I am repeatedly late for work and I have relied on the rooster's wake-up call so far, I eliminate the rooster—of course it can live on—and use an alarm clock.*
- *If I arrive too late despite the alarm clock, I adjust the alarm time.*
- *If the traffic jam on the motorway then repeatedly prevents me from arriving on time, then I circumvent this motorway with a new route or take the train.*

9.2.1 Creativity Techniques

Probably you and the experts in the team have already developed some suitable solution ideas in the course of the project. Therefore, it can be useful to combine the workshop for Root-Cause-Analysis with the search for solutions. Basic creativity and moderation techniques can support this process (Fig. 9.3).

[2]Video 33 (https://www.youtube.com/channel/UCShsy50l5dFsoSXU__hBKSA)

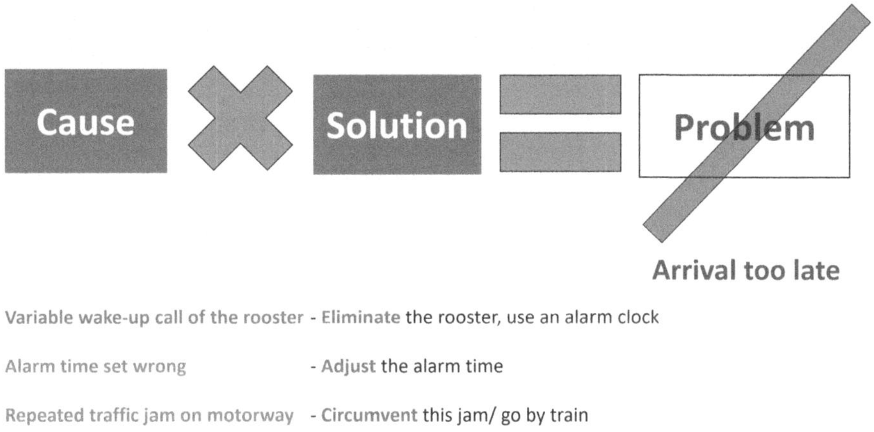

Fig. 9.2 Root causes (x') are the interfaces for solutions to eliminate, adjust, or circumvent a problem (Y)

Creativity techniques	
6-3-5 Method (Brainwriting)	6 participants note 3 ideas each on their worksheet. Then they pass it to the respective neighbours. They add their own 3 connotations. After moving 5 times the worksheets arrive at the source again. Result: 108 ideas in 30 min.
Six Thinking Hats	Separation of motives guiding cognitive processes into six consecutive phases, marked by different colours (hats for the particpants)
Morphological Analysis	New design variants for given objects, such as products and services, based on new combinations of the characteristics of their attributes
TRIZ	Problem solving method to overcome trade-offs or dilemma between contradictory objectives or elements

Fig. 9.3 Overview of selected creativity techniques

The **6-3-5 Method** is a structured brainstorming technique developed by Bernd Rohrbach (1968): 6 participants write down 3 ideas each on a worksheet, e.g., within 5 min. Then the worksheets are passed to the respective neighbors. They are inspired by the existing ideas on the received sheet and add their own 3 connotations within 5 min. After each sheet has been passed 5 times, it arrives at the respective source again. The result after 6 rounds in 30 min is 108 ideas (Fig. 9.4).

The biggest advantage is that each participant must develop his or her own ideas in the first round (= Brainwriting), before being inspired and guided by the results of other participants in the following rounds. Ideas are not immediately torpedoed, when they are expressed.

The **Six Thinking Hats**, developed by Edward De Bono (1986), are based on the assumption that problem-solving processes are guided by six divergent cognitive–emotional motives. De Bono recommends separating these motives into phases in a workshop, so that they can develop their specific strengths and do not interfere with

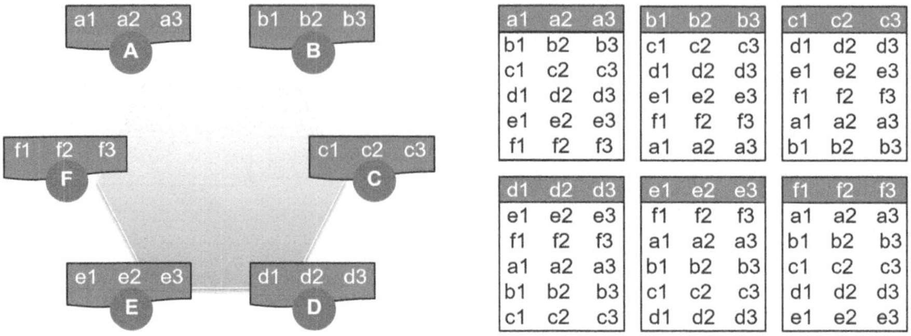

Fig. 9.4 6-3-5 Method: 6 participants write down 3 ideas each on a worksheet—e.g., within 5 min—and pass it 5 times. The result is 108 ideas in 30 min

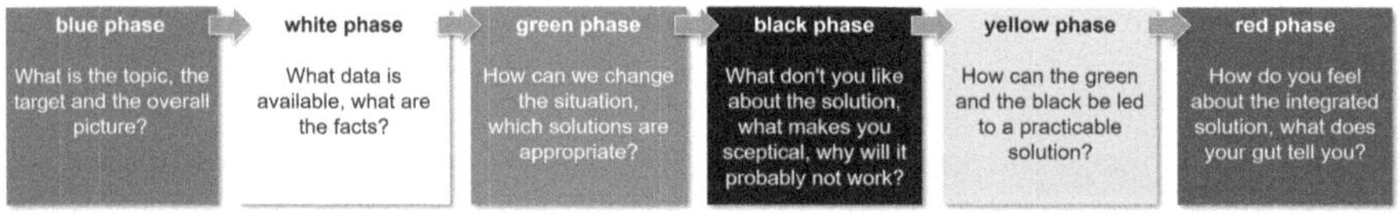

Fig. 9.5 Six Thinking Hats: Sequential phases for the creative development of solution ideas

each other. To visibly structure the process each of the phases is marked by a specific color. And in each phase only one of these colors should dominate (Fig. 9.5).

Blue What is the topic, the target, and the overall picture? This color defines the framework of the workshop and focuses the participants in our situation on the task of finding solutions to root causes.

White What data is available, what are the facts? This color represents the information that is already available from the Root-Cause-Analysis.

Green How can we change the situation, which solutions are appropriate? This color stands for creative solution finding. And this phase could be an interface for the 6-3-5 method.

Black What do not you like about the solution, what makes you skeptical, why will it probably not work? In this phase, well-founded arguments against the solution ideas are to be expressed.

Yellow How can the green and the black be led to a practicable solution? This color integrates the creative green and the pessimistic black.

Red How do you feel about the integrated solution, what does your gut tell you? This color represents emotional aspects and offers the possibility to rethink the solution or to accept it with a good feeling.

It is important that these phases are sequentially kept apart in time because people are good at criticizing and devaluing other people's contributions, often even before a sentence has been finished. This sometimes happens almost reflexively out of habit, without the person thinking explicitly: "Hey, that's a great idea. It has only one disadvantage: It's not mine." Suggestions are quickly relativized by objections, expressed in killer phrases and justifications. And an extraverted Black, e.g., nearly always beats an introverted Green. In the end the proposed idea is often forgotten, nothing is written down and they jump to the next idea, which could experience the same fate. In this case, it is all about power, about who will prevail—and not about creativity.

In situations like this, where there is a continuous back and forth between experts, I put on my virtual blue hat and ask: "Sorry, I've lost orientation. Is there something we should note on the flip chart?"—Always keep especially the green and the black phase strictly separate. First creativity, then realistic objections, and finally integrated, accepted solutions.

The **Morphological Analysis** by Fritz Zwicky (1969) seeks new design variants for given objects, such as products and services. For this purpose, all relevant attributes of the object and their conceivable characteristics are listed in a table (Fig. 9.6).

For example, if we want to create a new cookie, we first list all the relevant attributes that a cookie has, e.g., dough type, basic flavor, diameter, shape, and topping. Then, we break down the attributes into their possible variants. By new combinations of these characteristics, you could create an yeast dough, salty, with tomatoes topping. And you probably know these pizza cookies from your supermarket.

A sophisticated and elaborate method for problem-solving, to overcome a trade-off or dilemma between contradictory objectives or elements, is **TRIZ**, typically rendered as: Theory of inventive problem-solving. It was developed by the inventor Genrich Altshuller (2005). It takes a lot of expertise to use this method successfully.

Morphological Analysis				
Attributes	**Characteristics**			
dough type	yeast dough	shortcrust pastry	puff pastry	
basic flavour	sweet	salty	spicy	sour
diameter	5 cm	10 cm	15 cm	20 cm
shape	round	oval	rectangular	quadradratic
topping	chocolate	dragées	nuts	tomatoes

Fig. 9.6 Morphological Analysis: Combination of uncommon variants of the attributes of a particular product/service

Design of Experiments (DoE)				
		Amount of chocolate (x2)		
		80g (x2.1)	100g (x2.2)	120g (x2.3)
Type of chocolate (x1)	full milk (x1.1)	Y1 ratings Y2 costs		
	dark (x1.2)			

Fig. 9.7 Design of Experiments (DoE): Adjustment of the settings of influences (x) that lead to optimal results for one or more attributes of the output (Y)

Thus, I do not want to go into that any further and also just briefly mention the **Design of Experiments** (DoE). The DoE is a powerful method for adjusting causes, e.g., to calculate the settings of influence variables (x) with which the optimum result is obtained for one or more output variables (Y) (Fig. 9.7).

As a simple experiment we could, e.g., take the type of chocolate ($x1$) as an influence variable and differentiate here between: Full milk ($x1.1$) and dark chocolate ($x1.2$). As a second influencing factor, we could vary the amount of chocolate ($x2$) on the three levels: 80 g ($x2.1$), 100 g ($x2.2$), and 120 g ($x2.3$). As output variables (Y) we could have the taste evaluated ($Y1$) and the cost of each cookie ($Y2$) determined. An analysis of variance (ANOVA) would now make it possible to determine those settings for a new cookie type, within the scope of this design, with the optimal result for taste and costs.

In another experiment, you could determine the optimum baking temperature ($x2$) and duration ($x1$), in order to maximize the taste ($Y1$) and minimize energy costs ($Y2$).

9.2.2 Development of Solutions

For the typical solution search in a Six Sigma project, a combination of the 6-3-5 method and the Six Thinking Hats efficiently leads to effective results. For the workshop I draw a table on a flip chart.

In the Hierarchy-Tree of the Root-Cause-Analysis for the problem: Cookie Taste bad ($Y1$), in one branch we had identified these three causes as interfaces for solutions: *Chef has too few opportunities to nibble on chocolate* ($x1.2.3$), *Manager has restricted the consumption of chocolate* ($x1.2.3.1$), and *Too much chocolate is unhealthy for young Chefs* ($x1.2.3.1.1$) (Fig. 9.8).

Search for Solutions				
	Root-Causes (Branches of Hierarchy Tree)	Solution Ideas	Objection	Solution
Y_01 Problem: Cookies Taste bad	Chef has too few opportunities to nibble on chocolate (x1.2.3) Manager has restricted the consumption of chocolate (x1.2.3.1) Too much chocolate is unhealthy for young Chefs (x1.2.3.1.1)	Keep the Chef away from the bakery	Chef will protest	./.
		Checking the quantity of chocolate just before mixing it into the dough and, if necessary, compensating for the missing chocolate	Considerably higher effort, further waiting times expected at the scale	./.
		Weigh more chocolate in advance to compensate for the Chef's nibbling	Unclear how much more chocolate has to be weighed; the amount will vary if the Chef has different amounts of nibbles in each case	Chef may weigh herself an extra portion of chocolate; she is also informed about the importance of the ingredients for the taste of the cookies
	Information in recipe exceeds the Chef's capabilities (x1.2.1.1)	Translate standard recipes for children; detailed description of activities illustrated with images	Acceptance of standards are low if they specified externally and then prescribed	Chef documents the baking process for standard recipes with her own photos and drawings; Manager coaches the creation of this standard baking book for the Chef

Fig. 9.8 Develop solution ideas to eliminate, adjust, or circumvent the root causes, identify objections, and reformulate ideas to practicable solutions

So, we want to find a solution where our Chef does not eat more chocolate than before while preserving the integrity of the ingredients.

Now, I let the participants write down solutions for 10 min, with which the causes could be eliminated, adjusted, or circumvented (**green**). Then, the experts present their solution ideas, one after the other, and I write them in the second column of the table. Of course, you can also use stickers for this again. We got the ideas: (1) *Keep the Chef away from the bakery*, (2) *Checking the quantity of chocolate just before mixing it into the dough and, if necessary, compensating for the missing chocolate* and (3) *Weigh more chocolate in advance to compensate for the Chef's nibbling.*

After this green phase, the ideas in the **black** phase are now reasonably criticized: (1) *Chef will protest*, (2) *considerably higher effort, further waiting times expected at the scale* and (3) *unclear how much more chocolate has to be weighed; thus, the composition of dough will vary, if the Chef has different amounts of nibbles in each order.*

After a short discussion, one practicable solution is developed (**yellow**): *Chef may weigh herself an extra portion of chocolate and she will be also informed about the importance of the ingredients for the taste of the cookies.*

Then we go to the other branch and its root cause of the Hierarchy-Tree: *Information in recipe exceeds the Chef's capabilities* (*x*1.2.1.1). The idea for the solution (**green**) is proposed: *Translate standard recipes for children; detailed description of activities, illustrated with images*. Here the objection (**black**) is raised that: *The acceptance will be low, if standard is simply developed externally and then prescribed to the Chef.* So as final solution (**yellow**) results: *Chef documents the baking process for standard recipes with her own photos and drawings. And the manager coaches the creation of this standard baking book for our little Chef.*

We then repeat this procedure for the root causes of the remaining problems. And after the workshop I transfer the results into sigmaGuide (Fig. 9.9).

First, I search for the related problem and then calculate or estimate the quality costs per year resulting from the problem. This should be a justified proportion of the quality costs already specified in the Project-Definition. I estimate that the bad taste of the cookies costs us about 100 € per year.

Then, I transfer the root causes from each branch of the Hierarchy-Tree to one cell for the problem in question. Next, I estimate how strong the problem is determined by its root causes. That is about 75% and 5%, respectively, so for both root causes

Solutions

Rank	Kano-Category	Costs of the Problem/ Year:	Problem	Root-Causes	Cause determines the Problem to:	Sum of Determination	Solutions	Benefit	Effort	Rank (Effort/ Benefit)	Reduction of Problem-Costs
1	Must-Be	100,00 €	Y_01 | Problem: COOKIES (BAKED) TASTE CRUMBLY-BLAND	x1.2.3 Chef has too few opportunities to nibble on chocolate x1.2.3.1 Manager has restricted the consumption of chocolat x1.2.3.1.1 Too much chocolate is unhealthy for young Chefs	75%	80%	Chef may weigh herself an extra portion of chocolate; she is also informed about the importance of the ingredients for the taste of the cookies	8	4	1	75 €
				x1.2.1.1 Information in recipe exceeds the Chef's capabilities	5%		Chef documents the baking process for standard recipes with his own photos and drawings; Manager coaches the creation of this standard baking book for the Chef	4	10	2	5 €

- Select the problem
- Estimate the quality costs of the problem
- Transfer the root causes of the branches of the hierarchy tree
- Estimate the percentual determination of the problem by root causes.
- Transfer the identified solutions
- Evaluate the relative benefits and efforts of the solutions

The ranking and the reduced costs give you hints for the selection of appropriate solutions.

Fig. 9.9 Select the problem, transfer the root causes, the related solutions and evaluate the solutions

about 80%. This sum of determination indicates the portion, how strong all root causes of the branches together determine the problem.

Next, I transfer the identified solutions into the corresponding cells and estimate the benefit of the respective solutions, i.e., the contribution of each solution to eliminate, adjust, or circumvent the related root cause. And I do the same for the effort for each solution, i.e., the necessary consumption of resources (time/costs) to implement the solution.

The resulting effect reflects the product: Quality cost x determination of the problem. This means that the quality costs of the problem could be reduced by this amount if the root causes could really be controlled by the solution. Although the information at this stage is only a rough estimate, it can give you an initial assessment of the solutions.

Finally, the rank of a solution results from the effort/benefit ratio. The higher the benefit of a solution is, in relation to the effort to implement the solution, the higher is its ranking position.

9.2.3 Selection of Solutions

The results of this evaluation are summarized in the Chart Solution-Selection. It shows all solutions relative to their respective effort and benefit. The size of the bubbles corresponds to the effect of the solution, i.e., the reduced quality costs (Fig. 9.10).

Here you should focus on the solutions of the lower right triangle, where the benefit is higher than the necessary effort. Nevertheless, there may be reasons to choose solutions of the left upper half triangle as well. In our case, the advantages of the complex standard will probably pay off in the long term.

After the development and selection of suitable solutions, we will now derive concrete measures. We will identify and minimize their risks by countermeasures and assign responsibilities and deadlines.

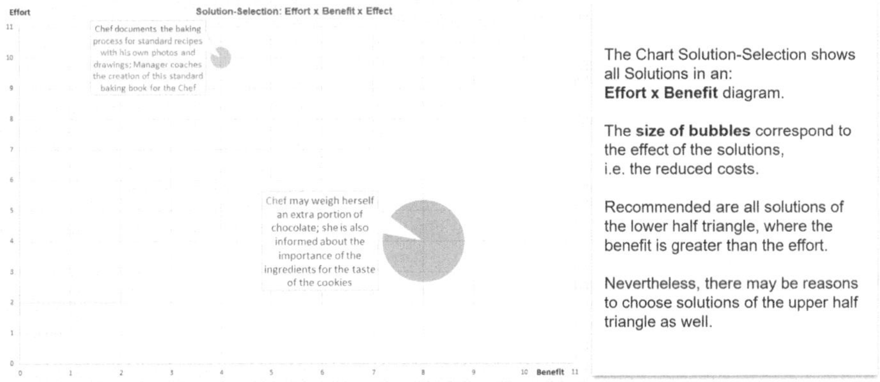

Fig. 9.10 Chart Solution-Selection: Select solutions where the benefit exceeds the effort

Questions (SC = Single Choice; MC = Multiple Choice)

Video 33/ Ch. 9.2/ #1 (MC)	Which of these methods support the elimination, circumvention or adjustment of root causes?					
		6-3-5 Method	Six Thinking Hats	Design of Experiments (DoE)	TRIZ	Morphological Analysis
1	Elimination					
2	Circumvention					
3	Adjustment					
3	points					

Tasks

Business-Projects: Please develop and select solutions to eliminate, circumvent, or adjust your root causes.

Standard-Project: Please develop and select solutions to eliminate, circumvent, or adjust your root causes. I think it is easy to develop solution ideas. Here are some general suggestions:

(a) Inform your interviewees about the harmfulness of littering for the environment and emphasize the responsibility of each individual to improve.
(b) Choose one of the areas where you collected data and clean up daily. The cleanliness of the place could reduce further littering.
(c) Wrap trash cans with paper in a color that makes them more visible.
(d) Draw a friendly, humorous signpost to the wrap paper of the trash cans.
(e) Write a short article about your project and the results so far for a local newspaper.
(f) Optimize the garbage emptying route together with the experts of the garbage removal service.

But probably you already have your own ideas.

9.3 Measures and Risks[3]

9.3.1 Action-Plan

An important cause of protracted and late implementation in project management is a lack of transparency, even of seemingly simple solutions. A realistic forecast of the implementation period and the probability of its actual implementation increases if you break down solutions into small work packages and put them in chronological order. And that is recommended with any project management. I am still often surprised by the number of specific tasks that result even from simple solutions.

[3]Video 34 (https://www.youtube.com/channel/UCShsy50l5dFsoSXU__hBKSA)

Action-Plan

Rank (Effort/ Benefit)	Reduction of Problem-Costs		Solutions	Measure-No.	Measure (What has to be done?)	Result (What will be achieved?)	Risk-Reduction-Measure (from FMEA)	Costs of implementation	Cost center	Deadline	Responsibility	Decision on implementation	Implementation-Status in %
1	75	€	Chef may weigh herself an extra portion of chocolate; she is also informed about the importance of the ingredients for the taste of the cookies	1.	1. Purchase of additional necessary chocolate 2. Let the chef weigh an additional 10g of chocolate for a nibble 3. Provide this extra portion on a separate plate 4. Chef is informed about the importance of the ingredients at every baking	Integrity of ingredients ensured	**Immediate brushing of teeth after each test of chocolate**	10,00 €	SISi123456789	15.11.20xx	Chef	yes	**60%**
2	5	€	Chef documents the baking process for standard recipes with his own photos and drawings; Manager coaches the creation of this standard baking book for the Chef	2.	1. Purchase of necessary resources for documentation (camera, notebook, etc.) 2. Introduction to operating the camera 3. Determining the activities in the process that are to be documented 4. Printout of photos in the office 5. Creation of the standard (photos, drawings, texts)	Notebook with step-by-step documented standard recipes	**Pilot the new standard based on one concrete recipe**	60,00 €	SISi123456789	31.12.20xx	Manager of Chef	yes	20%

Influences that can make implementation more difficult:

- Individual measures are too complex → Break them down into specific tasks and put them in their chronological order
- Expected result of implementation is unclear → Determine the expected result
- Diffuse responsibility for implementation → Determine responsibilities and deadlines
- Resources for implementation are necessary → Inform your Sponsor and let him/her decide
- Resistance to implementation → see Stakeholder-Communication
- Implemented measures are not accepted → Have the Sponsor present the measures to all affected employees before implementation
- Return to previous habits → Let the Sponsor check compliance with new standards after their implementation

Risks of the measures are reduced by:

- Carrying out an FMEA
- Exemplary implementation for a product/service on a line, at a location (piloting)

Decision for the implementation:

The decision for the implementation of the measures is made by the Sponsor (process owners) within the IMPROVE Steering presentation.

Prepare this presentation so well that decisions can be made at the end of the presentation if possible. It is useful to inform participants in advance about the measures to be taken.

Fig. 9.11 Action-Plan: Specify measures to implement the solutions and monitor their progress

On the left side of the Action-Plan in sigmaGuide, the developed solutions are listed with their effort/benefit rank and the expected reduction of problem costs, if the causes are under control. In the next columns the solutions can be specified as concrete measures (Fig. 9.11).

First break down the solution into concrete tasks and arrange the tasks in chronological order. For our taste problem we will: (1) *purchase additional chocolate*, (2) *let the chef weigh an additional 10 g of chocolate for her nibble*, and we will (3) *provide this extra portion on a separate plate*. Additionally, (4) *the chef is informed about the importance of the ingredients for the taste at every baking*.

Then we specify the result which should be achieved by the implementation: *Integrity of ingredients ensured*. The definition of the expected result enables all stakeholders to form an impression of the intended effect of the solution and later facilitates a before/after comparison.

The next column shows a risk reduction measure resulting from the FMEA. Since you can only give the next details in the Action-Plan, when you have completed an FMEA, I would like to briefly introduce this next tool of the sigmaGuide.

9.3.2 FMEA: Identification and Evaluation of the Risks of Measures

The FMEA, Failure-Mode and Effects-Analysis, is in itself a complete problem-solving method. It is often used as a stand-alone-tool in many technical areas to eliminate the risks of complete products and services or their components. The purpose of the FMEA in Six Sigma is to reduce the risks of measures before they are implemented. Thus, the measures are the starting point for us in the first columns of the FMEA (Fig. 9.12).

In our FMEA we want to identify:

- **Failures** or **problems**, which may result undesirably from the implemented measures
- Currently given **control options** to proactively detect these problems
- **Effects** of the problems on the business or the customer and the
- **Causes** of these problems.

After this qualitative analysis, the problem, its causes and effects are quantitatively evaluated.

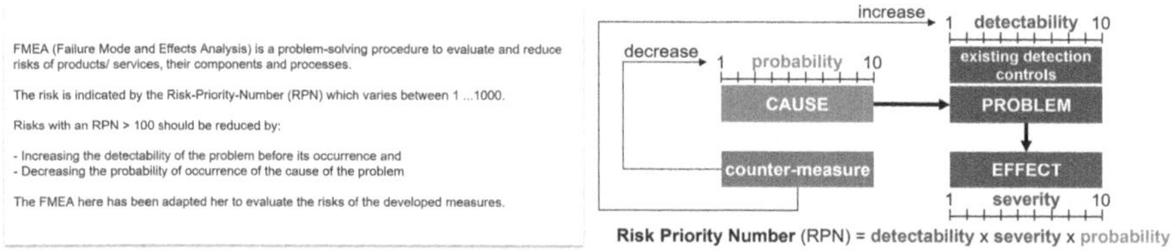

Fig. 9.12 FMEA: Specify the risks of measures and reduce critical risks by countermeasures

The experts assess how high the **proactive detectability** of a threatening problem **by the existing controls** is. This detectability is estimated on a scale of 1 to 10, meaning here:

- 1: high (!) detectability by the existing controls and
- 10: low (!) detectability.

Then the **severity of the effect** is evaluated on the same scale, meaning here:

- 1: low severity and
- 10: high severity of the effects on the business or the customer.

The **probability of occurrence of the cause** is also evaluated on a 10-level ordinal scale, meaning here:

- 1: low probability and
- 10: high probability of the cause, which leads to its respective problem.

The estimated values for the detectability of the problem, the severity of the effect, and the probability of occurrence of the cause are multiplied by each other. The result is the **Risk Priority Number (RPN)**, which can vary between 1 and 1000. If the risk exceeds an upper limit—I would like to recommend RPN > 100—then countermeasures should be defined.

Our FMEA is now about identifying possible failures or new problems that can result from the implementation of the specified measures. *We assume that the extra portion of chocolate can lead to caries for our Chef* (Fig. 9.13).

The actual control to proactively detect this problem is a: *periodic dental prophylaxis*. And we evaluate the detectability of the problem with a moderate value of 5 (1 = high, 10 = low detectability).

The potential effect of the problem could be a: *loss of teeth*. And for our young Chef we rate the severity of this effect with an 8 quite high.

The direct cause of the problem is: *bacterial plaque* and the probability of the cause is 10 because bacterial plaque always occurs when we eat chocolate with sugar.

By multiplying the detectability of the problem with the severity of the effect and the probability of the cause, a Risk-Priority-Number RPN = 400 results. This risk value exceeds the threshold of 100 and thus we need to develop countermeasures.

Because we cannot change the severity of the effect—a tooth loss remains a tooth loss—we focus on decreasing the probability of the triggering cause. The experts are convinced that: *immediate brushing of teeth after each nibble* would be an appropriate countermeasure.

Based on our countermeasure, we conduct a second risk analysis. Now we estimate 2 as a new value for the probability of bacterial plaque. An improvement of the detectability does not seem to be necessary to us, because the twofold prophylaxis per year is generally recommended. The new resulting RPN = 80 and lies below the limit of RPN = 100, so that we can finish the risk analysis for this solution here.

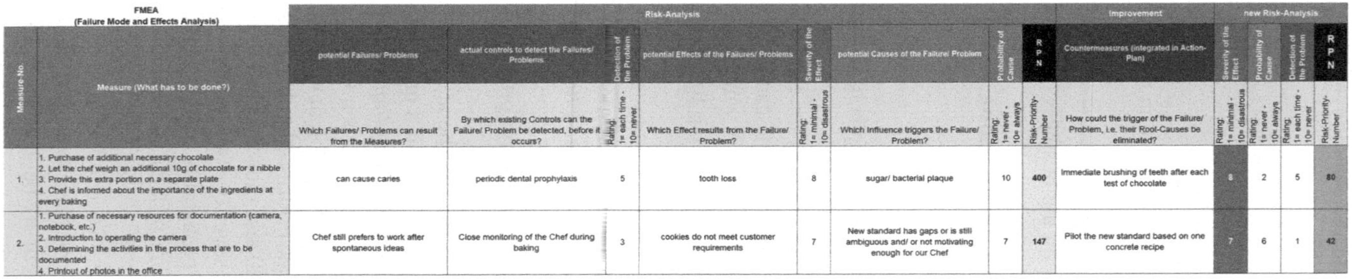

FMEA (Failure Mode and Effects Analysis)

Measure-No.	Measure (What has to be done?)	Risk-Analysis: potential Failures/ Problems	actual controls to detect the Failures/ Problems	Detection of the Problem	potential Effects of the Failures/ Problems	Severity of the Effect	potential Causes of the Failure/ Problem	Probability of Cause	RPN	Improvement: Countermeasures (integrated in Action-Plan)	new Risk-Analysis: Severity of the Effect	Probability of Cause	Detection of the Problem	RPN
		Which Failures/ Problems can result from the Measures?	By which existing Controls can the Failure/ Problem be detected, before it occurs?	Rating: 1= each time - 10= never	Which Effect results from the Failure/ Problem?	Rating: 1= minimal - 10= disastrous	Which Influence triggers the Failure/ Problem?	Rating: 1= never - 10= always	Risk-Priority-Number	How could the trigger of the Failure/ Problem, i.e. their Root-Causes be eliminated?	Rating: 1= minimal - 10= disastrous	Rating: 1= never - 10= always	Rating: 1= each time - 10= never	Risk-Priority-Number
1.	1. Purchase of additional necessary chocolate 2. Let the chef weigh an additional 10g of chocolate for a nibble 3. Provide this extra portion on a separate plate 4. Chef is informed about the importance of the ingredients at every baking	can cause caries	periodic dental prophylaxis	5	tooth loss	8	sugar/ bacterial plaque	10	400	Immediate brushing of teeth after each test of chocolate	8	2	5	80
2.	1. Purchase of necessary resources for documentation (camera, notebook, etc.) 2. Introduction to operating the camera 3. Determining the activities in the process that are to be documented 4. Printout of photos in the office	Chef still prefers to work after spontaneous ideas	Close monitoring of the Chef during baking	3	cookies do not meet customer requirements	7	New standard has gaps or is still ambiguous and/ or not motivating enough for our Chef	7	147	Pilot the new standard based on one concrete recipe	7	6	1	42

Fig. 9.13 FMEA in sigmaGuide: Specify the risks of measures and reduce critical risks by countermeasures. Countermeasures are backward linked with the Action-Plan (see Fig. 9.11)

The countermeasure is backward linked with the Action-Plan (see Fig. 9.11). Thus, we will additionally implement it together with the already specified measures.

Now we can estimate the costs of implementation. To secure financing, we propose a cost center to be used for the billing of expenses. In the last steps we suggest a responsible person for the measure and determine a deadline to make the implementation binding.

For our second solution, we define a *concrete procedure for creating our standard recipes*, leading to a *notebook with step-by-step documented standard recipes*, including a *training of the new standard.*

With these measures to introduce a new process guiding method, we see a great risk that our Chef still prefers to work according to her spontaneous ideas and that we cannot meet the customer's requirements. We do not see any individual, targeted, and effective countermeasures here and therefore decide to initially test the new standard in a **pilot application**, based on one selected recipe. Because before we completely convert the bakery to our Chef's new recipe book, we first want to see if it works in principle. If not, we can immediately fall back on the established process for the next orders.

This concludes the most important preparations for the IMPROVE steering presentation. This is because the decision, for or against implementation, is made by the Sponsor and the process owners.

If there are important changes in your process flow, if activities are dropped, new activities added, if the order of the activities is to be changed, then you should **map the new process** to increase understanding and acceptance. Assign changed methods and resources to activities, as you did in the MEASURE phase, but now for the new should-be process.

Prepare this presentation so well that decisions can be made at the end of the session if possible. Additionally, it is useful to inform the Sponsor and process owners in advance about the suggested measures, so that they are neither surprised by the measures nor by the need to decide on them.

In case of positive decisions, please monitor the progress of implementation, then you are not surprised if the implementation is delayed.

There are still two general risks during and after implementation:

1. The measures are not accepted by all affected co-workers. We cannot always be certain of the actual acceptance before we do. As a rule, the Sponsor should therefore proactively demonstrate his/her commitment by presenting the measures to all affected employees prior to implementation.
2. Affected employees return to their previous habits after a short period of time. To proactively prevent this, please have the Sponsor announce that he or she will continuously check compliance with the new standard after its implementation.

If this does not help, you must find out the causes of the resistance. Do the employees know all the reasons why the solutions were introduced? Do they really have the necessary competence and support to apply the new process? Do they perhaps have considerably more work now? The same principles apply here as in stakeholder analysis and communication (Sect. 6.4). However, if you do not see a chance for a sustainable implementation, then please ask your Sponsor for further help.

With the implementation of the measures the IMPROVE phase ends. Now we are ready to check our improvements and to further control the process in the CONTROL phase.

Questions (SC = Single Choice; MC = Multiple Choice)

Video 34/ Ch. 9.3/ #1 (SC)	Please assign the evaluation categories to the risk components of the FMEA.			
		Detectability	Severity	Probability
1	Cause			
2	Problem/ Defect			
3	Effect			
3	points			

Tasks

Business-Projects: Please derive concrete measures from the selected solutions and prevent risks with countermeasures from an FMEA. Copy the results of all IMPROVE tools and, as far as possible, pictures from the implemented measures in your Project-Story-Book and comment on the results.

After implementation, schedule at least 1 month for data collection on the new process or status to perform a statistically reliable before and after comparison. For a Chi^2 Test and a *t*-Test you should collect at least 5 new data per condition. Strong improvements can be significant from this sample size.

Standard-Project: Please derive concrete measures from the selected solutions and prevent risks with countermeasures from an FMEA. Copy the results of all IMPROVE tools and pictures from the implemented measures in your Project-Story-Book and comment on the results.

I know that the implementation will be difficult. And the sustainability of your success will not be guaranteed. But for this standard project, you do not just achieve our TUM Green Belt certification. Your expertise, courage, and commitment can be a strong reference for your first or next employer.

After implementation, schedule at least 1 month for data collection on the new process or status to perform a statistically reliable before and after comparison. For a Chi^2 Test and a *t*-Test you should collect at least 5 new data per condition. Strong improvements can be significant from this sample size.

References

Altshuller, G. (2005). *40 Principles: Extended Edition*. Worcester, MA: Technical Innovation Center Inc.

De Bono, E. (1986). *Six thinking hats*. London: Penguin Life.

Rohrbach, B. (1968). Kreativ nach Regeln—Methode 635, eine neue Technik zum Lösen von Problemen. *Absatzwirtschaft*, *12*(19) (1 Oktober 1969).

Zwicky, F. (1969). *Discovery, invention, research—Through the morphological approach*. Toronto: The Macmillan Company.

10 CONTROL

10.1 Overview: Steps, Tools, Their Purpose, and Results[1]

In the CONTROL phase we can hopefully reap the fruits of our labor. We will focus primarily on comparing our measurands before and after implementation of our measures, on their financial benefits, and on ensuring sustainability.

First, we examine the new data with descriptive graphical analyses, e.g., Pareto-Chart, Boxplot, and Time-Series-Plot. We then monitor the process performance over time with Control-Charts, e.g., the I-MR-Chart and calculate the process capability, e.g., Pp/Ppk and the Sigma Level (Fig. 10.1).

The CONTROL phase will focus on a statistical review of the impact of our measures, which hopefully will show that the improvements are statistically significant and practically relevant. To do this, we use difference hypotheses for the important output variables. Based on significant differences, we can also often reliably calculate the financial benefit and at least estimate the nonfinancial benefit. And then compare your results with our targets from the Project-Charter.

To ensure the sustainability of the improved process performance, we create a Process-Management-Plan and establish a Continuous Improvement Process (CIP) to avoid regressions and to further improve in small steps.

This brings us to the end. It is good style if you summarize your lessons learned and point out further improvement potentials that you have identified during the project. Finalize your Project-Story-Book, prepare for the final steering and presentation with all participants, and hand over the project including its future duties (Process-Management-Plan; CIP) to the process owners.

It is not much anymore. The collection of the new data could become laborious again, but your goal is within reach.

[1] Video 35 (https://www.youtube.com/channel/UCShsy50l5dFsoSXU__hBKSA).

© Springer Nature Switzerland AG 2019

R. Hutwelker, *Six Sigma Green Belt Certification Project*, Management for Professionals, https://doi.org/10.1007/978-3-030-31915-1_10

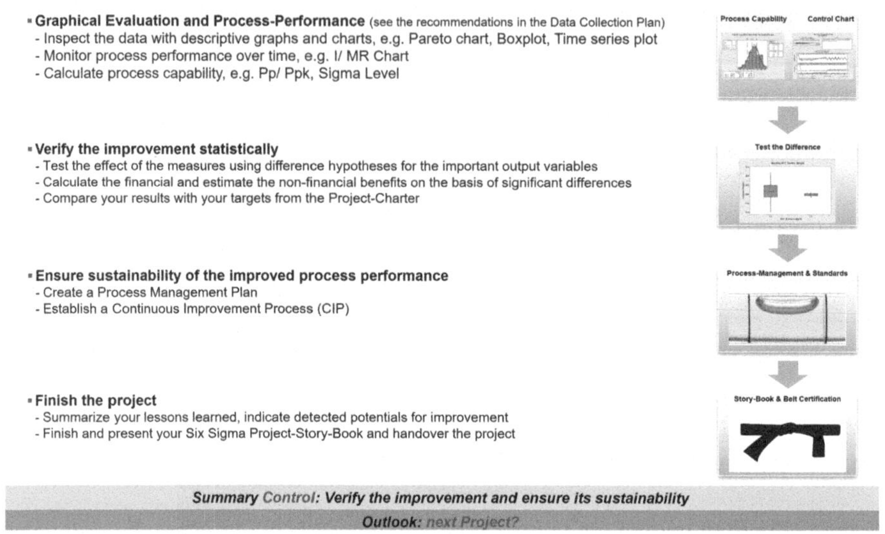

Fig. 10.1 Overview CONTROL: From the evaluation of the improvement to sustainable success

10.2 Data-Evaluation[2]

To examine the data, you need of course new data from the time after the improvement. You already know from the ANALYSE phase how to check the data for plausibility and evaluate it graphically and with descriptive statistics (Mean, Standard Deviation).

We collected data for 4 weeks after improvement (Fig. 10.2). First, I transformed the new rating data again. Values above and equal to the specification limit: 4 are assigned to the category: ok and values below the rating: 4 to the category: defect (see Fig. 8.14).

The comparison of the frequencies of all problems shows that the overall number of defect cookies was reduced by 38% ($N = 100$). Primarily the number of defects in taste—83% reduction—and weight of ingredients—80% reduction—were affected. Thus, it came to a shift of the ranking of the error frequencies. The *wrong cookie weight* now comes first and the *wrong cookie diameter* second. But there are still about 10% *bad tasting cookies.*

The Boxplot compares the rating data before improvement with data from the first week after improvement. Here it is shown that the mean taste even increased from: xbar = 2.46 (before) to xbar = 3.92 (after first week) (1 = very good; 6 = very bad), meaning that the taste decreased. The dispersion of taste also increased from: $s = 1.20$ (before) to $s = 2.47$ (after first week).

[2]Video 36 (https://www.youtube.com/channel/UCShsy50l5dFsoSXU__hBKSA).

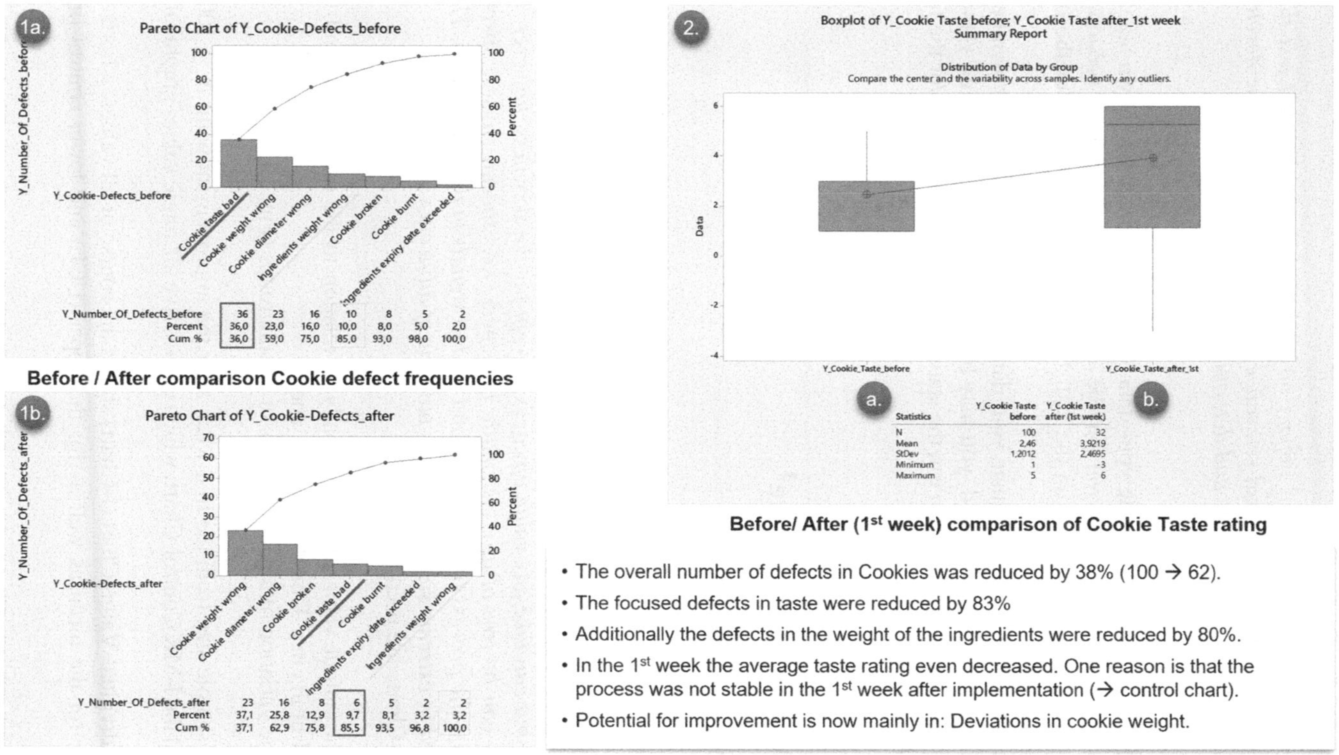

Fig. 10.2 The overall number of defect cookies was reduced by 38%. In the first week, however, the taste ratings even decreased from xbar = 2.46 to xbar = 3.92 (1 = very good, 6 = very bad)

Before the improvement the ratings scattered from 1 = very good to 5 = bad. After the improvement the ratings ranged from 1 to 6 = very bad. And there is at least one invalid value (−3).

Our measures have had a real impact on the new sample. We will test their statistical significance and practical relevance later. First, we look at the development of the new process over time and determine its process capability.

Tasks

Business-Projects: Please evaluate your data, e.g., with Minitab, as shown here. If your Minitab license has expired, you may be able to install a new demo version on another notebook. You can also run the statistical tests with *Excel* and/or the free programs *R* or *gretl*.

Standard-Project: Please evaluate your data, e.g., with Minitab, as shown here. If your Minitab license has expired, you may be able to install a new demo version on another notebook. You can also run the statistical tests with *Excel* and/or the free programs *R* or *gretl*.

10.3 Process Performance[3]

10.3.1 Process Control

In a controlled process, only random influences, i.e., common causes, occur. These random influences lead to normally distributed data of the output, if the data were measured on a cardinal scale. The distribution of data of cycle times can be steep on the left and skew on the right. This is because the cycle times cannot fall below a lower limit but can take any value upwards. Except for the skewness, however, the cycle times can be free of any systematic influences, i.e., special causes. Specific influences on the process on the other hand lead to signals in the output data. They are reflected in extreme single values or patterns of adjacent values.

Control-Charts are used to:

(a) Map the output values chronologically in relation to the mean value and the control limits in a Time-Series-Plot, and to
(b) Map the variation of adjacent values or in subgroups (Fig. 10.3).

For our sample size from the first week after the improvement with: $N \leq 100$, I recommend the I-MR Control-Chart, which can resolve all single values graphically:

(a) The Individual-Value-Chart (I-Chart) shows the course of all values, a horizontal line for the mean as well as for the upper (UCL) and lower control limit

[3]Video 37 (https://www.youtube.com/channel/UCShsy50l5dFsoSXU__hBKSA).

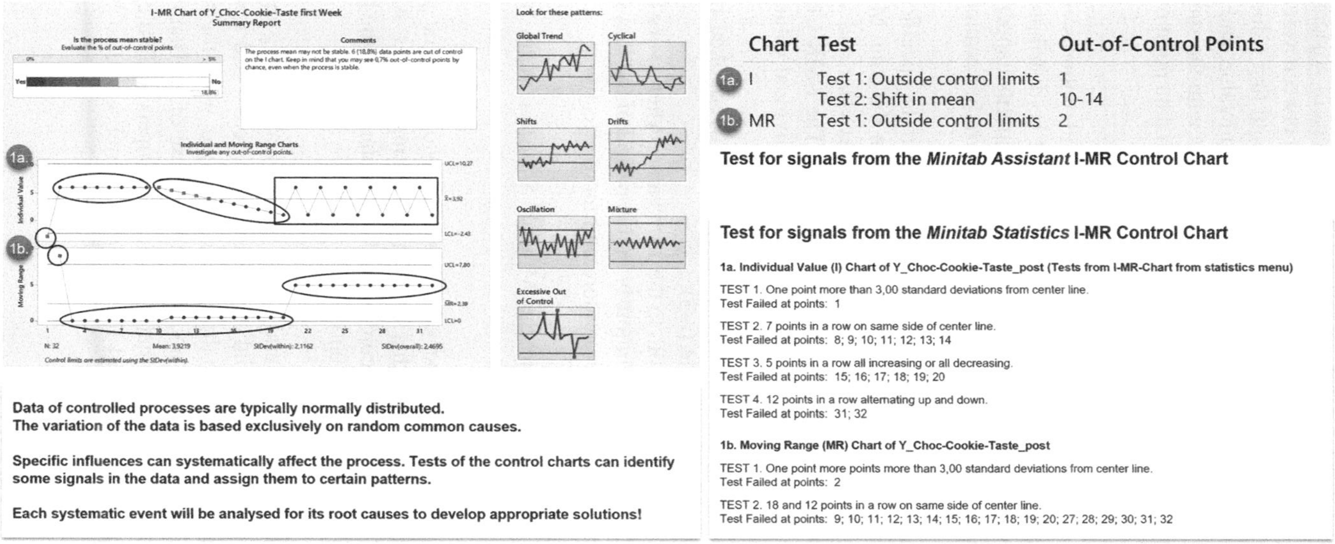

Fig. 10.3 Patterns in the data of the first week after implementation indicate that the process was not yet under control and needs to be analyzed for specific causes

(LCL). In the xbar(-S)- and xbar(-R)-Chart the sample is divided into subgroups and each data point in the chart represents the mean value of a subgroup.

(b) In the Moving-Range-Chart (MR-Chart), the differences of adjacent values are mapped. Here the data points indicate the variation between values. The first data point corresponds to the difference of the second and first value, the second data point refers to the difference of the third and second value, and so on. In the (xbar-)S-Chart the variation is indicated by data points with the standard deviation (s) in the subgroups and in the (xbar-)R-Chart each data point represents the range (R) in the respective subgroup.

Our data of the first week show extreme values and patterns, conspicuous deviations from a random scattering and therefore cannot be distributed normally. Statistically it is possible that such extreme values and patterns happen by chance, but the probability for this is extremely low. We, therefore, rather assume underlying special causes.

In addition to graphically displaying the values and limits, specific tests of the Control-Charts can identify extreme values and patterns in the data. Minitab offers two variants of the mentioned Control-Charts: via the *Assistant* and via *Statistics* menu, which differ in the extent of these tests on signals. The I-MR-Chart from the *Assistant* displays two test results for the course of the individual values:

Test 1: One value outside the control limits. This is the overlooked invalid value outside the range of the rating scale (−3).

Test 2: A shift of the mean from the 10th to the 14th value. After a very bad start, the taste of the cookies improved continuously in this period of the first week. In this ramp-up phase, our measures became more and more effective.

The tests of the I-MR-Chart from the Minitab *Statistics* menu index further patterns:

TEST 2 Indicates a suspicious number of values in a row on the same side of center line. This indicates that we started with a series of bad tasting cookies. As a cause we assume that our Chef did not work according to the new standard at the beginning of the week and continued to nibble on chocolate.

TEST 4 Indicates that there are several points in a row alternating up and down. This ping-pong can point to sequential dependencies, where a bad result causes a good one and vice versa. This may be the case, for example, when the response of a process to an intervention is delayed and we enter a phase of alternating over- and understeering. But it is also possible that our Chef was supervised on every second order, applying the new standard and then returning unobserved to her previous habits.

For the Moving-Range-Chart (MR-Chart), the

TEST 1 also indicates an extreme value in both Control-Charts. Here, the large difference between the invalid value and the subsequent valid value is displayed as an outlier (see TEST 1 of Individual-Values-Chart, I-Chart).

TEST 2 points on 18 and 12 points in a row on same side of center line. This means that in the first period, the difference between all adjacent values was relatively small and that the differences between adjacent values in the second period was relatively big. The causes for this are the continuous bad cookies in the first period and the ping-pong with big differences between adjacent values in the second period.

To bring the process under control, each systematic event should be analyzed, and its root causes should be eliminated. This strategy may become cumbersome, but the result can be a controlled excellent process with only common causes leading to only random variations. And this procedure already describes the future monitoring of the process for extreme values and patterns (Process-Management-Plan).

10.3.2 Process Capability

For the calculation of the process capability now, the ramp-up data of the first week will be excluded. This section also repeats the corresponding evaluation from the ANALYSIS phase, only with new data and thus allows a before–after comparison. I select the *Before/After Capability Analysis* of the Mintab *Assistant* (Fig. 10.4).

The specified target as well as the specification limits are of course the same for this comparison.

The statistical parameters show improvements in the mean and standard deviations. In relation to the target and specification limits this also leads to an improvement in the capability indices: Ppk and Z.Bench, our Sigma Level, which increased from 1.93 to 4.90. The corresponding percentage outside the specification dropped from 2.66% to 0.00%, i.e., a difference of 2.66%, which is indicated as a 100% reduction.

This evaluation of the Minitab *Assistant* additionally displays statistical tests of the position and dispersion of the process. The change in dispersion is tested by the 2-Sample Standard Deviation Test, i.e., the F-Test.

The chart for the F-Test indicates that the standard deviation of the ratings is significantly reduced in the new sample ($p < 0.001$). And the next diagram also shows, based on the 2-Sample t-Test, that the mean value of the ratings has improved significantly from 2.46 to 1.65. And this improvement is also practically relevant as we think. We now offer our customers a better and more consistent taste experience.

The Histograms for the before–after comparison with the overlaid target and the upper specification limit (USL) demonstrates these changes. The related Anderson–Darling Normality Test indicates a significant deviation from the normal distribution. Thus, again the results must be interpreted under reserve.

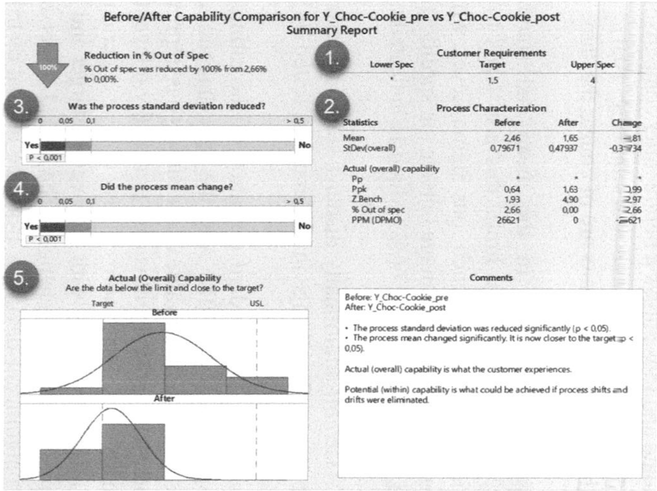

- **Significant and practically relevant increase of the taste judgement**, from 2.46 to 1.65.
- The **variability of taste** was also significantly **reduced**.

→ We offer customers a better and more consistent taste experience.

- This improvement is reflected in the process capability, increasing approximately 3 Sigma Levels, from 1.93 → 4.9 Sigma.
- The performance of the process is now closer to the goal (1.5). **(see below: 1-Sample t-test for difference to target)**

The Process Capability Analysis shows:

1. The customer requirements are defined as specification limits on the rating scale of taste:
   - Upper specification limit (USL): 4 **(see Data-Collection-Plan)**
     Taste should be rated better than 4:= adequate
   - Target: 1,5 **(see Data-Collection-Plan)**
     The mean of taste rating should be ≤1,5:= very good/ good)

2. Improvement of statistical parameters:
   xbar: 2,46 → 1,65 - Difference: 0,81
   s: 0,80 → 0,48 - Difference: 0,32

   Improvement of capability indices:
   Ppk: 0,64 → 1,63 - Difference: 0,99
   Z.Bench: 1,93 → 4,90 - Difference: 2,97
   % Out of spec: 2,66 → 0,00 - Difference: 2,66 (100%)

3. The standard deviation is significantly reduced (p< 0,001)
   (2 Sample Standard Deviation Test/ F-Test)

4. The difference in the means: is significant (p< 0,001)
   (2 Samples t-Test)

5. Histograms for the Before vs. After comparison:
   The Anderson-Darling normality test shows a significant result.**

** **interpretation of results under reserve**

Fig. 10.4 The before vs. after comparison of the process capability shows a significant improvement. Focus is now on the statistical significance and practical relevance of the improvement

But we can see that we clearly improved our chocolate cookies. The ratings of taste increased in the sample 0.81 rating points on a 6-level rating scale. And regarding customer requirements, we have improved the taste of the chocolate cookies by about 3 Sigma Levels (Z.Bench: from 1.93 → to 4.9).

Although it makes sense that the Minitab Assistant for process capability already contains statistical tests, this is not the case in every statistics program. Therefore, I will now formulate the appropriate hypotheses and calculate these tests again separately.

Questions (SC = Single Choice; MC = Multiple Choice)

Video 37/ Ch. 10.3/ #1 (MC)	What do the tests of control-charts indicate?	
1	Outlier	
2	Shifts	
3	Drifts	
4	Relationships	
5	Cycles	
6	Differences	
7	Oscillations	
8	Significances	
9	Trends	
10	Capabilities	
10	points	

Video 37/ Ch. 10.3/ #2 (MC)	Please assign the the necessary requirements to the following capability indices and control-charts.			
		Data normal distributed	N ≤ 8	N > 8
1	I-MR-Chart			
2	Cp/ Cpk			
3	Pp/ Ppk			
4	xbar-S-Chart			
5	xbar-R-Chart			
5	points			

Video 37/ Ch. 10.3/ #3 (SC)	What do these capability indices evaluate?			
		Cpk/ Ppk/ Z.bench	Defect %/ Z.bench	DPU
1	Tolerance/ standard deviation			
2	Defects outside specification			
3	Defects per Unit			
3	points			

Tasks

Business-Projects: Please prepare the necessary Control-Charts and calculate the process capability for your most important Ys.

Standard-Project: Please prepare the necessary Control-Charts and calculate the process capability for your most important Ys.

10.4 Improvements and Benefits[4]

10.4.1 Statistical Tests of Improvement

You probably remember that the two hypotheses on differences and relationships depend on the content of your assumption and on the scale level of your variables. Do you have an idea what hypotheses we need to formulate to verify the improvement in any Six Sigma project?

The standard hypothesis for the CONTROL phase is: There is a/no difference in: Taste (Y) between: the states of x_before vs. x_after. And we will test this hypothesis first for the rating of taste (Y) with the t-Test to differentiate between the two nominally scaled states.

Now that you can read these results, you will recognize that this 2-Sample t-Test for independent samples leads to the same result, as the included t-Test of the capability analysis: The new cookies taste 0.81 grades better than the old ones and this result is statistically significant. Thus, we can reject the H0 and adopt the HA: Our improvement is statistically confirmed. And we are convinced that the size of the difference also has a practical benefit for our customers and thus for our bakery (Fig. 10.5).

With the next t-Test we want to again evaluate the degree of how good the target 1.5 (see Project-Charter) for the chocolate cookies was achieved (Fig. 10.6).

And we formulate the corresponding hypothesis: There is a/no difference in: Taste (Y) between: the rating and the target (Y'). Here we use the 1-Sample t-Test, because we compare a sample with a criterion value. And, contrary to the above hypothesis, where we wanted to find a significant difference between before and after, we here wish that the H0 is confirmed, as in the test for normal distribution, because that would mean that the target value is within the confidence interval of the sample, meaning that there is no difference between the data and the target value.

But the result is again significant ($p = 0.001$). The mean of the taste ratings is: xbar = 1.65 and its confidence interval (CI) ranges from 1.57 to 1.73. The target value 1.5 is not included and thus there is a significant difference between the taste ratings and the target. Although we improved the taste of our cookies, we did not reach our target. And we do not know what caused it yet. I will mention this in the final presentation of my project and will neither justify nor make excuses for it. For although we have not achieved our target, we can demonstrate concrete benefits from our improvements.

10.4.2 Financial and Other Benefits

Hopefully you also got nice results in your project. Now, it is about deriving the financial and other benefits and presenting the overall result plausibly. There is no

[4]Video 38 (https://www.youtube.com/channel/UCShsy50l5dFsoSXU__hBKSA).

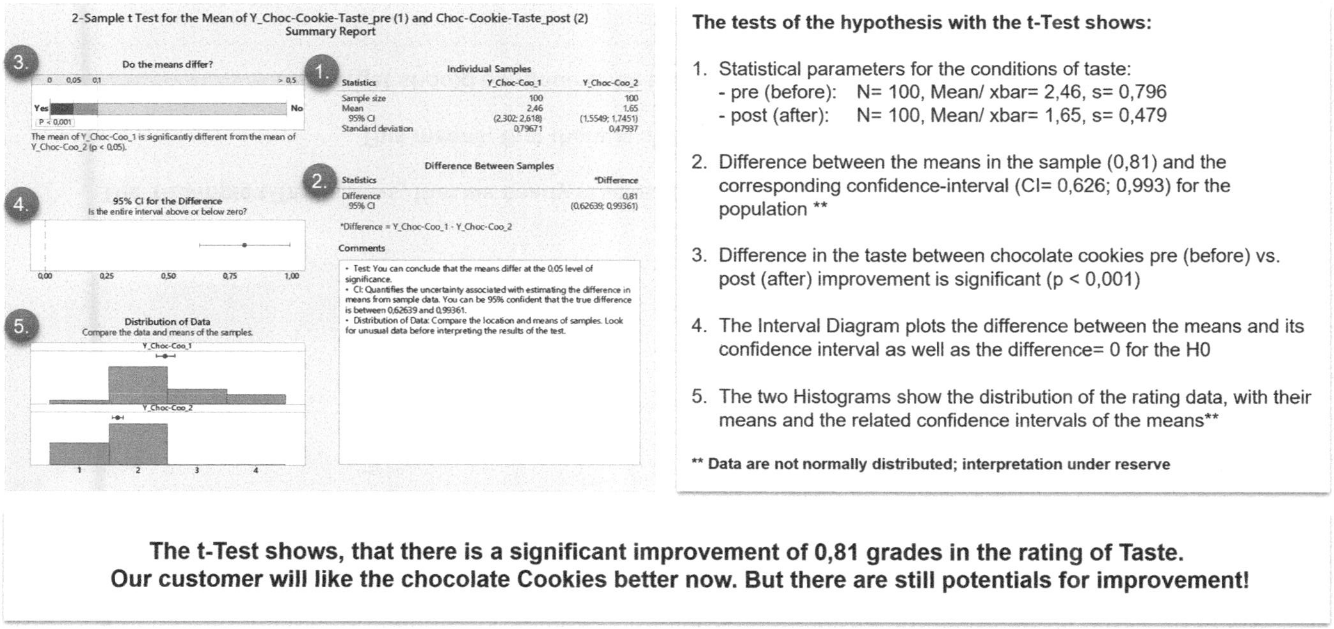

Fig. 10.5 Hypothesis: There is a/no difference in: taste (Y) between: the states before vs. after improvement (x). Focus is now on the degree how good the target for the chocolate cookies was achieved

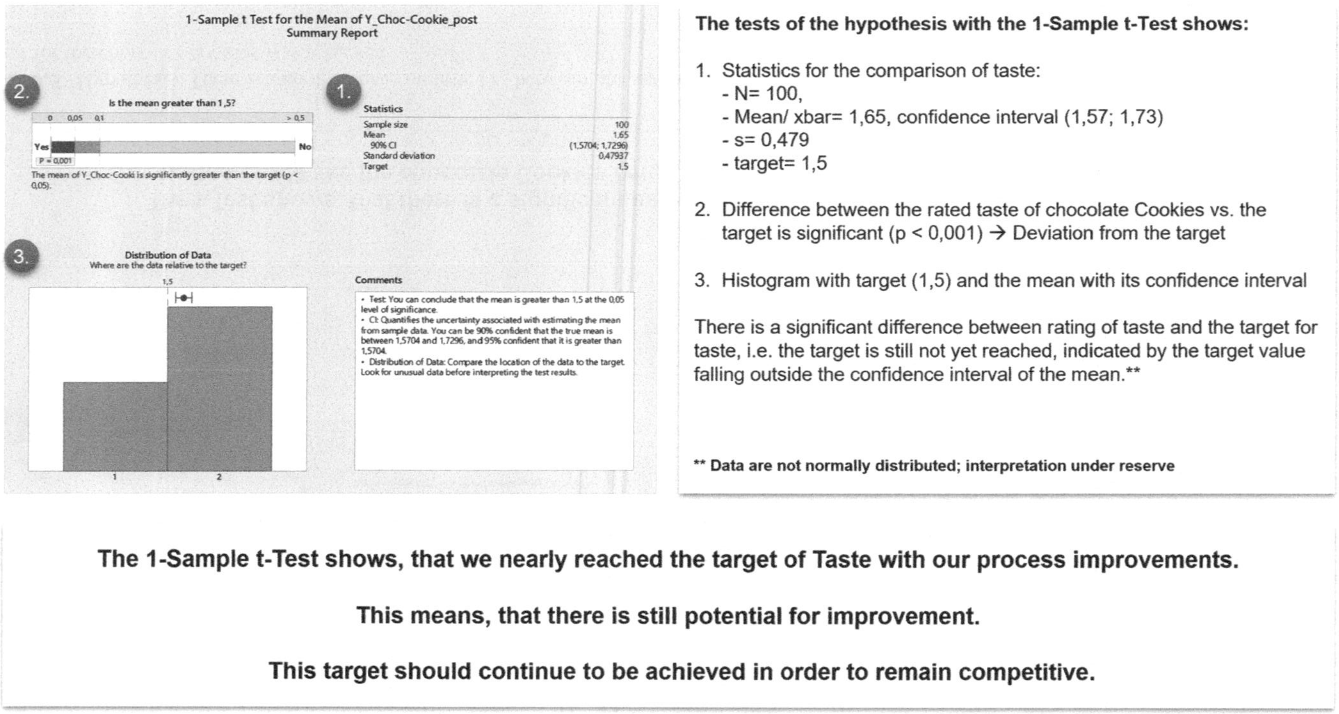

Fig. 10.6 Hypothesis: There is a/no difference in: taste (Y) between: the ratings and the target (Y'). After all before–after comparisons tested, we will now deduce the benefits of the improvements

standard procedure for this, but I would like to give a few clues. The starting points for the results overview are the defined problems, the underlying root causes we have detected, and the associated measures that we have implemented.

For cardinally scaled data, we use the *t*-Test or Analysis of Variance (ANOVA) to verify the improvement. Here we can translate the mean difference of significant results into financial benefits. So, we might ask ourselves how much the difference has a financial impact. The mean value will represent our expected benefit (Fig. 10.7).

But no serious Belt will be satisfied with a point estimate. We can additionally use the lower value of the confidence interval to give an estimation of the worst case and the upper value of the confidence interval for an estimation of the best case. Based on these confidence limits, we could, e.g., now calculate how much more cookies we can sell in the worst and best case in the future and derive the profit from this.

Furthermore, we can transform cardinally scaled data down to nominal scale level (see Fig. 8.14), or we can use originally nominally scaled data to detect changes in the defect rate (%) and evaluate them in monetary terms. The basis for this could be the reduced percentage of defective cookies outside the specification (see Process Capability). And the 2-Sample % Defective test offers the statistical verification of our defect reduction. And then you can again use the upper and lower limits of the confidence interval to estimate the best and worst case.

To illustrate the results for more than one problem, you can use a Pareto-Chart to show the reduced quality costs for the defined problems and compare them with the Pareto-Chart of the original quality costs. We therefore expect that the overall reduction in quality costs will be higher (256 €) than previously estimated in the solutions section (80 €) (Sect. 9.2) (Fig. 10.8).

You may need the help of your financial controller to calculate and confirm the financial benefit. Of course, it can happen that your statistically tested results are doubted, e.g., with the indication of a possibly wrong sampling or a measuring error. Here you can refer to your Data-Collection-Plan and to the Measurement-System-Analysis (MSA), if you have performed one. The argument of a possibly too small sample size can of course be countered by the fact that your tests have led to a significant result for exactly this sample size. But if you consult the financial controller before the calculation, then the later acceptance by the controller and by your Sponsor will increase.

In deriving the other nonfinancial benefits, you may be able to mention the standard methods that have been introduced, which are likely to lead to better, homogeneous, and stable process performance in the long term.

This way you can fill the table (Fig. 10.7) with all your solved quality, availability, and consumption problems.

You may have been able to implement Quick Wins already during the project. This is your achievement, and therefore, these Quick Wins should also be listed in this table. Always make sure that you fairly appreciate the contributions of all those involved in the project, so that no one envies and relativizes your success.

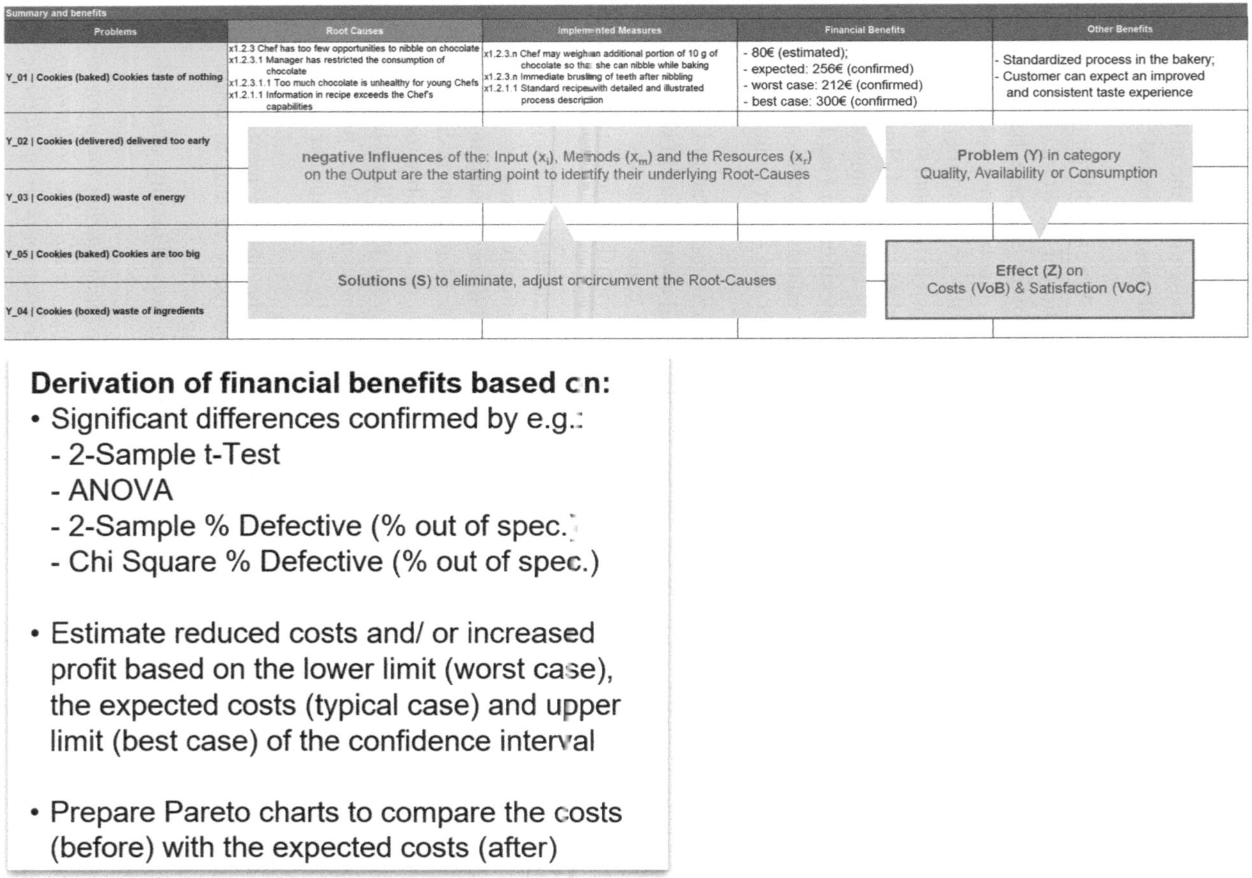

Fig. 10.7 Summarize the main results—from the problems to the measures implemented. Assign the financial/other benefits to the measures and have them independently confirmed

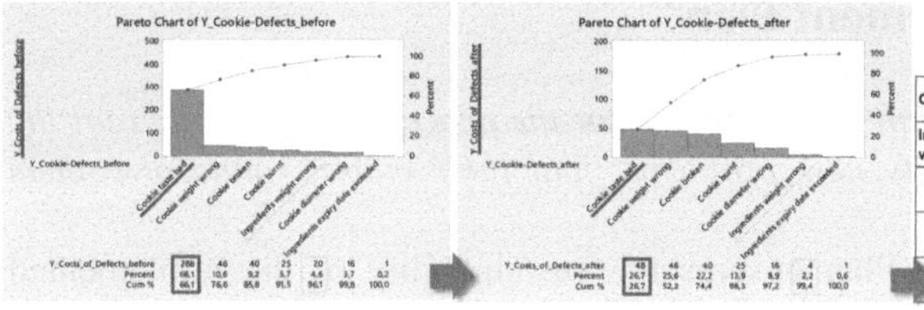

	Costs (before)	Costs (after)	Savings
Cookie taste bad	288,00 €	48,00 €	240,00 €
Ingredients weight wrong	20,00 €	4,00 €	16,00 €
		expected	256,00 €
		worst case	212,48 €
		best case	299,52 €

Fig. 10.8 Assign the financial/other benefits to the measures and have them independently confirmed

Questions (SC = Single Choice; MC = Multiple Choice)

Video 38/ Ch. 10.4/ #1 (MC)	Which tests are suitable for a before-after comparison?	
1	2-Sample-t-Test	
2	ANOVA	
3	Chi-Squared-Test	
4	Regression-Analysis	
5	Correlation	
6	2-Sample-%-Defective-Test	
6	points	

Video 38/ Ch. 10.4/ #2 (SC)	Which result from a statistical test should be the basis for aconservative estimate of the financial benefit?	
1	Lower limit of the confidence interval	
2	Expected value (e.g. the mean difference)	
3	Upper limit of the confidence interval	
3	points	

Tasks

Business-Projects: Please verify your improvements statistically and derive the financial and other benefits.

Standard-Project: Please verify your improvements statistically and derive the financial and other benefits. It is difficult to derive financial benefits from a change in attitude in the questionnaire replies. But you could, for example, calculate what a halving of the effort of manually cleaning the trash cans environment would mean.

10.5 Process-Management-Plan[5]

The Process-Management-Plan is a document for the process owner to monitor the process in the future, identify new problems, find their causes, and solve them (Fig. 10.9).

The Process-Management-Plan ensures the sustainability of the implemented measures by:

(a) Monitoring of the future process performance with Control-Charts and
(b) By responses to performance drops determined in a reaction plan.

The starting point in sigmaGuide is again the defined problems and their measurement, operationalized already in the Data-Collection-Plan. In the ANALYSIS and CONTROL phase so far, we wanted to get a first impression of the performance of the process before and immediately after improvement. And we were interested in extreme values and patterns in the data. Now, it is about controlling the processes in the future—possibly without your participation—so using the Control-Chart as a diagnostic tool for the employees and the process owner. Let me suggest a simple strategy for developing this Control-Chart (Fig. 10.10).

First, I choose an appropriate time interval for the regular updating of the Control-Chart to monitor the taste of our cookies. You can schedule the interval: *weekly*, *bi-weekly*, *monthly*, or *quarterly*. I prefer close monitoring, at least at the beginning. That is why I am planning to let our customers test the taste weekly, with a sample size of 25 cookies. The 25 data can also be displayed very clearly as individual values in a Control-Chart and therefore I get the proposal: I-MR-Chart. I personally think that up to 100 values can be resolved by an I-MR-Chart. The I-MR-Chart is also my personal favorite, but of course it does not always fit. Depending on the scale level, sample size, and the number of displayed data points, the xbar-R-, xbar-S-, p-, and u-Charts are also suggested (Fig. 10.11).

In addition to the Control-Chart type, we must now determine the control limits and the center line. In contrast to the defined specification limits, the control limits are calculated based on the standard deviation or the moving range. Each new value and of course a high volatility in the data then always leads to new and, therefore, to probably inappropriate control limits. That is why I would like to use proven reference values in the future.

If we start with a new process, then we must estimate the control limits and the center line from the actually underlying sample. For future monitoring, however, it is important to derive reference values from a controlled process phase.

In the example I measured data over two weeks. In the menu to the preliminary Control-Chart I first choose: *Estimate from the data* (1a). Even before the chart is

[5]Video 39 (https://www.youtube.com/channel/UCShsy50l5dFsoSXU__hBKSA).

The **Process-Management-Plan** ensures the sustainability of the implemented measures by:

a. Monitoring of the future process performance with **control charts** for the critical outputs, with determined values for their control limits (UCL, LCL) and center line and a schedule for their actualisation

b. Reactions to performance drops with a **reaction plan**, which determines, how signals in the control charts are to be treated by whom **(additional document, not part of sigmaGuide)**

- The results of the control charts should be discussed regularly, daily or weekly in the first weeks after implementation, in order to identify weak points immediately and react to them
- Important part of the reaction plan are regular meetings of a Continuous Improvement Team, which analyses the causes of the signals and develops solutions
- The process owner should know and accept the reaction plan and place the control charts visibly in his office. A copy should be visible to all employees on a blackboard (Visual Management)

Fig. 10.9 Purpose and implications of a Process-Management-Plan

	Process-Management-Plan	Define measures to sustainably maintain the process-improvements							
Ranking of Outputs (Y)	Outputs (Y)	Measurand	Unit	Target and specification limits (USL; LSL)	Scale-Level	In which time intervals will the control chart be actualized?	How large will the sample size be in each time interval?	How many data points should the control chart represent?	I-MR Chart (if N <= 100)
Output (Y)		Data from Data-Collection-Plan							
1	Y_01 | Problem: COOKIES (BAKED) TASTE CRUMBLY-BLAND	rating-scale	grades 1 .. 6	Target: 1.5 LSL: 4	Data Rank Ordered (Ordinal-Scale)	weekly	25	25	25 data points; no subgrouping

Fig. 10.10 Determination of the suitable Control-Chart depending on the sample size and the data points to be represented

displayed, the menu already immediately shows the identified signals and allows them to be selectively excluded from the calculation (1b). If signals are already displayed in the dialog, then exclude at least the extreme values. This is because these outliers determine the standard deviation disproportionately and thus extend your control corridor disproportionately.

The result for the first week shows the already known course (see Fig. 10.3). And the excluded outliers are indexed as signals here (1b). In the second week, the data show a smooth controlled course (2a). I would like to make this second, controlled period the basis of my reference value calculations. Therefore, I copy the data of this controlled period into a new variable and let for a new Control-Chart again estimate the limits and center line from these data. If necessary, I exclude the now identified extreme values, calculated for this period, again. This new Control-Chart now only shows the controlled course of the second week (2a).

So that this controlled period can serve as a reference for the future course, I enter the calculated values for the limits and center lines as known values in the table of the menu for the final Control-Chart (3a). For examination I generate this chart for the data from both periods. Compared to the first Control-Chart, you can see that almost all values of the first uncontrolled week are now outside the control limits. And the data of the controlled period are now within the control corridor in the same way as with the second Control-Chart. Now we have created an ambitious reference for checking future data (3b) (Fig. 10.12).

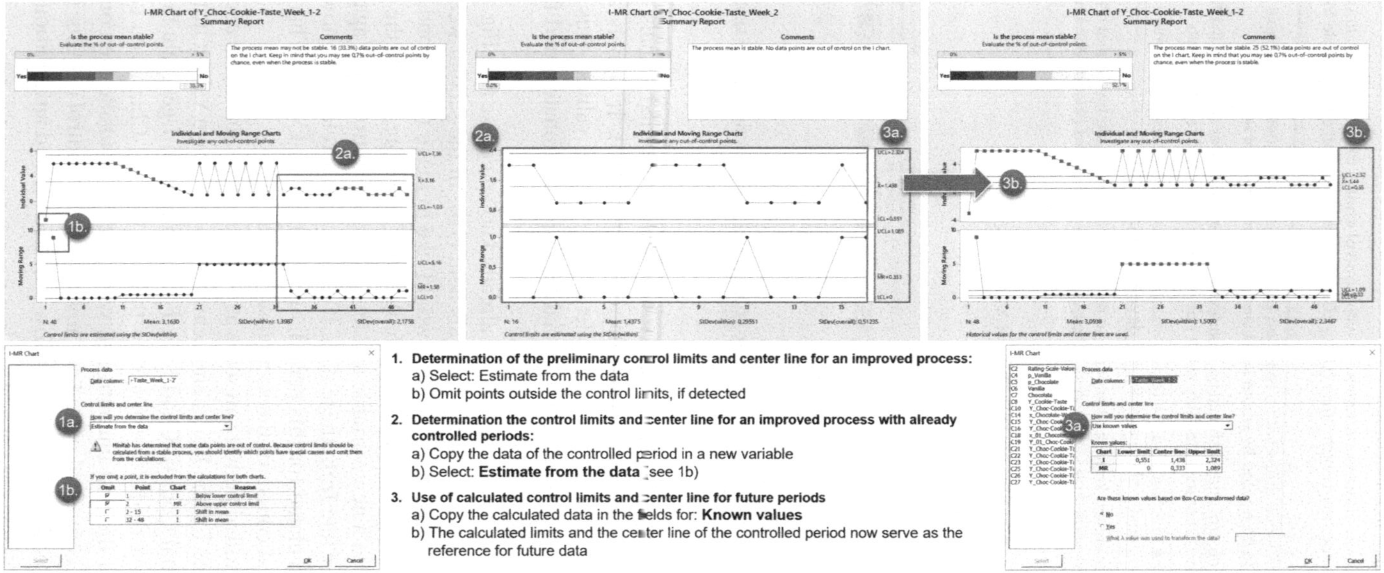

Fig. 10.11 Determination of appropriate control limits for Control-Charts. For the first try: Estimate from the data—For sustainable monitoring: Use known values from controlled period

I-MR Chart (if N <= 100)	xbar-R Chart (if N > 100 and if subgroup size <= 8)	xbar-S Chart (if N > 100 and if subgroup size > 8)	p-Chart (if ok vs. ko is discriminated)	u-Chart (if ok vs. different defect opportunities are discriminated)	Which control limits should be used? (LCL; Center-Line; UCL)	Who is responsible for creating the control charts?	In which document is the reaction plan specified?	Who is responsible for maintaining the reaction plan?
25 data points; no subgrouping					**I-Chart:** UCL: 2.324 Center-Line: 1.438 LCL: 0.551 **MR-Chart:** UCL: 1.089 Center-Line: 0.333	Mrs X	Reaction-Plan.xlsx	Mr Y

Fig. 10.12 Enter the calculated control limits (LCL, UCL) and the mean value for the center line, serving as a basis for future process control plus responsible persons for creating and maintaining the Control-Charts

Y_Cookie-Taste	UCL	Center Line	LCL
-3	2,324	1,438	0,551
6	2,324	1,438	0,551
6	2,324	1,438	0,551
6	2,324	1,438	0,551
6	2,324	1,438	0,551
6	2,324	1,438	0,551
6	2,324	1,438	0,551
6	2,324	1,438	0,551
6	2,324	1,438	0,551
6	2,324	1,438	0,551
5,5	2,324	1,438	0,551
5	2,324	1,438	0,551
4,5	2,324	1,438	0,551
4	2,324	1,438	0,551
3,5	2,324	1,438	0,551
3	2,324	1,438	0,551

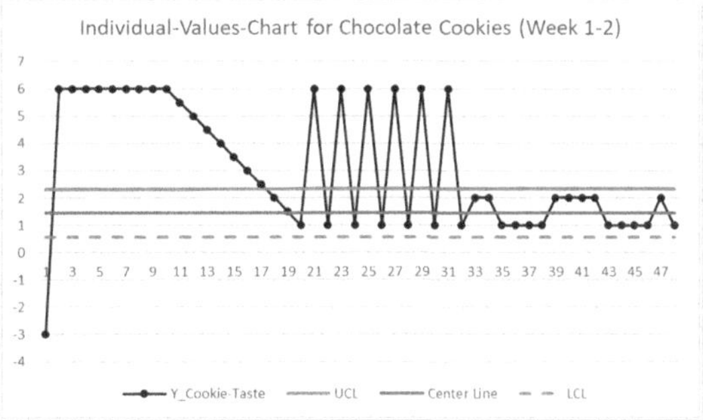

Fig. 10.13 Reproduction of a simple Individual-Value-Chart (I-Chart) with Excel

These calculated values for the control limits and the center line can now be entered in the Process-Management-Plan. Finally, you can specify who is responsible for maintaining the Control-Chart and the reaction plan and enter the filename of the reaction plan.

For specific signals in the Control-Charts a reaction plan now determines: *What must be done by whom?* This reaction plan is not part of sigmaGuide and should be created according to your company standard.

Of course, the process owner and his employees must be informed about the use, maintenance, and above all, about the potential signals in Control-Charts (see Fig. 10.3). The installation of a Continuous Improvement Team (CIP) would be ideal if there is not yet one, in which the results of the Control-Charts are discussed weekly, extreme values and patterns are examined for causes, and suitable solutions are developed. I would like to recommend that you accompany this team for a while so that they get used to the work with its results and indications. To support the team, you could easily rebuild simple Control-Charts, e.g., with Excel (Fig. 10.13).

Maybe your Sponsor is even willing to build a process dashboard with the Control-Charts and their updates. In any case, these dashboards should always be in view, one for the staff and one for the process owner. This would have introduced or added visual management to your department.

Questions (SC = Single Choice; MC = Multiple Choice)

Video 39/ Ch. 10.5/ #1 (SC)	In which sequence can ambitious control limits for checking future data in a new process be determined?					
		1. Step	2. Step	3. Step	4. Step	5. Step
1	Estimate control limits from the subsample					
2	Exclude (at least) extreme values from calculation of LCL/ UCL					
3	Extract a subsamble of the new measurement data that is/ seems under control					
4	Create a control-chart for the complete sample of new measured data					
6	Take the calculated LCL/ UCL as reference values for the final control-chart					
6	points					

Tasks

Business-Projects: Please create a Process-Management-Plan and prepare at least one Control-Chart for future monitoring of the process.

Standard-Project: Please create a Process-Management-Plan and prepare at least one Control-Chart for future monitoring of the process.

11 Project Completion

11.1 Review and Outlook

The last task is not mandatory, but it demonstrates your good style until the end. The review is about your: Lessons learned and the outlook about: New and unsolved potentials that you have identified during the project.

It may not be easy to list the lessons learned in this table because you have been on the road with the project for a long time and maybe you do some things quite naturally differently than before. To make it easier for you to look back, I have listed a few key points, according to which you can structure your learnings, concerning the subject matter and product, the process, methods and tools, people and team, management, finance and the company.

It has been a long and laborious journey to achieve the results. And you can be proud of all the things you can additionally enter here. They are part of your personal transformation process (Fig. 11.1).

The second part is about your outlook. Here you can enter all unsolved problems you have discovered during the project. Perhaps they contain the topic for your next project, or it becomes the topic of a Green Belt project for a colleague. Both would help to strengthen the entire Six Sigma program in your company.

11.2 Checks and Evaluation

When you get here, you will have made it. The last tool in sigmaGuide is a checklist for you and a grading scale for your Master Black Belt. Please check for all tools whether you have done them partially, completely or not and whether you have

Video 40 (https://www.youtube.com/channel/UCShsy50l5dFsoSXU__hBKSA)

© Springer Nature Switzerland AG 2019

R. Hutwelker, *Six Sigma Green Belt Certification Project*, Management for Professionals, https://doi.org/10.1007/978-3-030-31915-1_11

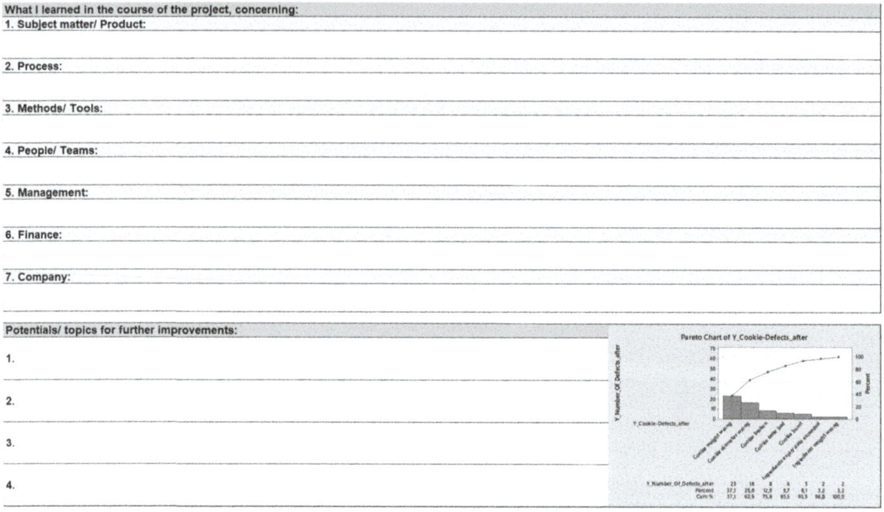

What I learned in the course of the project, concerning:

1. Subject matter/ Product:
2. Process:
3. Methods/ Tools:
4. People/ Teams:
5. Management:
6. Finance:
7. Company:

Potentials/ topics for further improvements:

1.
2.
3.
4.

Fig. 11.1 Lessons learned and potentials for further improvement topics

documented them in your Project-Story-Book. If something is missing, please complete it as far as possible (Fig. 11.2).

After project completion, a Master Black Belt will evaluate the quality of your results. Here you can achieve 0–10 points per task. If you receive at least 5 points in all tasks, your project will be certified (50%). If you earn more than this 50%, you will be credited for your Black Belt certification if you want to go for this next level. With the best result—10 points in each task—you will receive a credit of 50%. In this case your Black Belt project does not have to be as demanding and complex as it would be without credits.

Tasks

Business-Projects: Please summarize your lessons learned. List the newly identified problems and their potential to stimulate further improvements. Check the application and documentation of all DMAIC tools and steps. Copy the results of all CONTROL tools in your Project-Story-Book and comment on the results.

Standard-Project: Please summarize your lessons learned. List the newly identified problems and their potential to stimulate further improvements. Check the application and documentation of all DMAIC tools and steps. Copy the results of all CONTROL tools in your Project-Story-Book and comment on the results.

11.3 End of Project

This is the end of our Six Sigma course. Let me thank you for your participation and for your patience in Six Sigma language.

Phase	Tool	Business-Project	Standard-Project	Your check	MBB grade
General	Project-Story-Book	All headers and footers of the slides are specified, results described, interpretations provided, conclusions drawn		yes	5
Define	Part 1	Identify a topic for a Six Sigma project			
	Project-Topic	Identify Problems of the daily work as a potential for Improvement		yes	7
	Part 2	Define a Six Sigma project			
	Project-Definition	Summary: Process, Output, Problem and Effect		yes	8
Define	Part 3	Implement a Six Sigma project			
	SIPOC	Structure the Process in its important steps, with related Supplier, Inputs and Outputs and Customer		yes	10
	Voice to Critical	Interviews conducted with customers, the sponsor and managers	Interviews conducted with representatives of the municipality/ garbage collector	yes	10
	Project-Charter	Complete and sign the Project-Charter		yes	10
	Stakeholder-Communication	Identify the Stakeholder and develop a Communication-Plan		yes	10
Measure	Input-Analysis	Describe Inputs (xI) of the Process, Requirements on the Inputs and Deficiencies		yes	10
	Process-Mapping and -Analysis	Workshop conducted & documented (photos: Process-Mapping & team))	If possible: Workshop conducted & documented (photos: Process-Mapping & team (municipality/ garbage removal))	yes	10
	C&E Matrix xY	Evaluate relationships between negative Influences of the Inputs (xI) and the Activities of the Process (xP) on the Outputs (Y)		yes	10
	Data-Collection-Plan	Operationalise Measurands of Inputs (xI), Activities (xP) and Outputs (Y); Recommendation for appropriate: Charts, Parameter, Process-Capability-Indices, Control-Charts, One-Sample-Tests		yes	10
	Hypotheses	Overview of all automatically generated Hypotheses, prioritized by their Risk; Recommendation for appropriate statistical Tests		yes	10
Analyse	Data Evaluation	Plausibility of data checked and data graphically displayed		yes	10
	Process Performance	Process performance determined, with capability indices and control charts		yes	10
	Hypotheses Tests	Hypotheses statistically tested		yes	10
	Root-Cause-Analysis	Workshop conducted & documented (photos: Root-Cause-Analysis & team))	If possible: Workshop conducted & documented (photos: Root-Cause-Analysis & team (municipality/ garbage removal))	yes	10
Improve	Solutions	Develop Solutions to eliminate, adjust or circumvent the Root-Causes		yes	10
	Action-Plan	Specify Measures to implement the Solution-Ideas		yes	10
	FMEA	Analyse the Risks of Measures (FMEA:= Failure Mode and Effects Analysis)		yes	10
	Implementation	Measures implemented (at least one for each problem)		yes	10
Control	Data Evaluation	Plausibility of data checked and data graphically displayed		yes	10
	Process Performance	Process performance determined, with capability indices and control charts		yes	10
	Hypotheses Tests	Hypotheses statistically tested		yes	10
	Process-Management-Plan	Define measures to sustainably maintain the process-improvements		yes	10
	Summary and Benefits	Summarize the results of the phases and demonstrate the financial and other benefits of the project		yes	10
General	Lessons Learned	Lessons learned summarized and topics identified for future projects		yes	10
	Evaluation:	96%	excellently passed (46% Black Belt Bonus)		

Fig. 11.2 Review of the tool application and its documentation by you and evaluation of the entire project by your Master Black Belt

I am glad you joined the course.

I hope that my activities carried out by my resources and guided by my methods could transform you into such an output that you are satisfied - because the knowledge, the practical competence, and the motivation you acquired are within your tolerance corridor, concerning quality, processing time, and the resources you consumed.

If you think that your yield is below your specification limit or if I even missed a *must-be*, then please let me know. And feel free to also give feedback if you reached your targets or if you even detected a *delighter* in this course.

Six Sigma Project Guideline

12

No matter if you are planning a Six Sigma project or if you have already completed it, this summary of the tools in the DMAIC phases might serve as a guideline for your project or to stabilize your already acquired knowledge.

There are two columns in this guideline (see Fig. 12.1), with the sequence of DMAIC phases: DEFINE and MEASURE on the left and ANALYSE, IMPROVE, and CONTROL on the right. All blocks in the columns correspond to just one tool.

On the right side of each tool you find traffic lights showing you who should participate in this tool. The legend on the bottom of the guideline assigns the different colors: **Red** for the management, **yellow** for the Green or Black Belt, and **green** for the team.

The bullets in the explanations of each tool show you whether the sigmaGuide or a statistical program, e.g., Minitab, supports the execution of the corresponding DMAIC tool.

The **DEFINE** phase starts with the identification of a **Project-Topic**. (Example: In our bakery, customers complain about the taste of our cookies). Not all challenges in a company are suitable for Six Sigma. At first you should identify a weakness in your processes, i.e., a weakness of your products and services. If a process is repeated, if there are more than about ten products or services generated or delivered per month, if the products/services show a weakness in quality, availability, or consumption, if an internal/external customer is affected by this problem and if the causes of the problems are suspected within the company's processes, then the topic is probably suitable for a Six Sigma project.

After the Project-Topic has been identified, it should be further defined (**Project-Definition**). In the first step, the process and its output are specified. Then the problems are defined as deviations of the output in terms of: Quality, Availability (quantity/time), and Consumption (input/resources) (Example: Cookie taste bad). Next the effects of the problems on the company and the customer are rated

Video 41 (https://www.youtube.com/channel/UCShsy50l5dFsoSXU__hBKSA)

© Springer Nature Switzerland AG 2019

R. Hutwelker, *Six Sigma Green Belt Certification Project*, Management for Professionals, https://doi.org/10.1007/978-3-030-31915-1_12

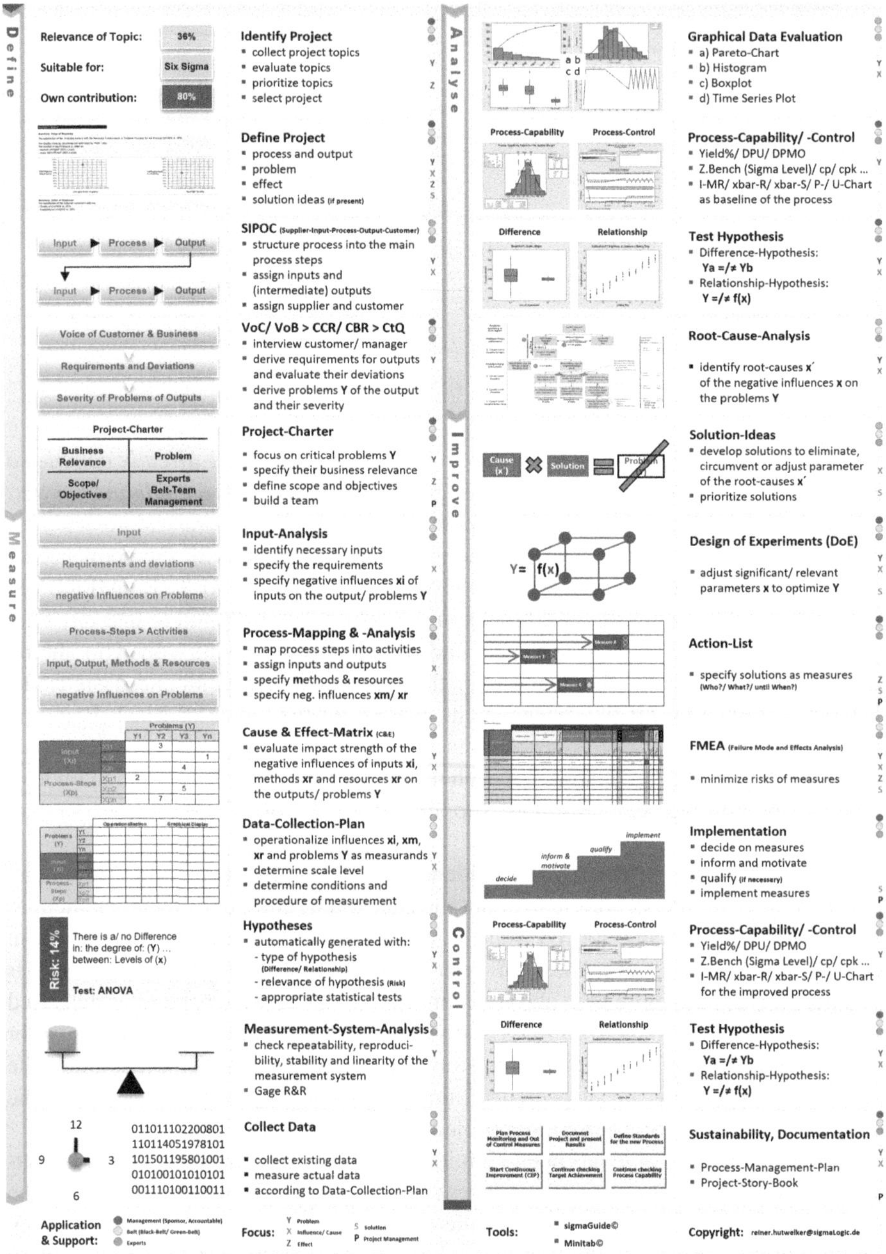

Fig. 12.1 Six Sigma Project Guideline presenting the chronology of tools in the DMAIC phases

according to given data. The quality costs of the problems for the company and further financial losses should be estimated here. This is the base to calculate the potential benefits of the project. Then the existing ideas for the solution of the problems can be collected, to later acknowledge already existing good ideas and to reduce resistance to the solutions in the IMPROVE phase. This information is summarized in the Project-Definition. And the Project-Definition might support the management in their decision which of the defined projects could be implemented by a Green or Black Belt with the support of a Sponsor from the management.

Now the project can start with the **SIPOC** (Supplier–Input–Process–Output–Customer). This tool draws the borders of the process field with the first relevant input, which might already cause the problem and the last relevant output, which has the problem (Example: *Y*_01: Cookies taste bad). The process flow between the first input and last output is structured into the main process steps. Each of these process steps leads to an intermediate output. All these intermediate outputs might also be associated with preliminary problems. Thus, it is necessary to identify the customers of all these outputs who are affected by their problems.

The next tool of the DEFINE phase: **Voice-to-Criticals** continues with the outputs of the SIPOC and their customer. Here it is necessary to ask the customer for their positive and negative feedback to these outputs. These "Voices of Customer" are translated into critical requirements (CCR), and then into problems, which are the deviations from the requirements. A classification of these requirements according to the Kano-Model and the rating how well they are fulfilled leads to the severe problems, the critical to quality (CtQ) (Example: Y_01: Cookies taste bad (*must-be*; 20% fulfilled; CTQ)).

The important information is now summarized in the **Project-Charter**, like the business case from the Project-Definition, the problems of the outputs and their effects on the customer and on the company. Objectives of the project are formulated according to the defined problems. Further specifications about the scope of the project: "What is in and what is out?" specifies the duties for the Green or Black Belt and assures the expected results for the management and the Sponsor.

After the necessary experts of the project team are determined, as well as the deadlines for the further DMAIC phases, the **MEASURE** phase can be started.

While the primary purpose of the DEFINE phase is to identify the *Y*s, i.e., the problems of the outputs, the purpose of the MEASURE phase is to specify the negative influences (x) on the output, which trigger the problems (Y). These xs, in the input and the activities of the process, in the used methods and resources, together with hypotheses about their relationship, serve to solve the central equation of Six Sigma: $Y = f(x)$.

First the external inputs are analyzed (**Input-Analysis**), listed already in the SIPOC. The requirements of the process on these inputs are evaluated here, as well as their deficiencies, which might have a negative impact on the outputs and trigger its problems (Example: xinput_01: Ingredients expiry date exceeded (1%)).

Next, the major process steps of the SIPOC are mapped in more detailed activities (**Process-Mapping**). It is a good idea to perform this mapping together with the experts of the process and to just follow and observe, as far as possible, the generation of a certain product or service along its value chain. For each activity, the guiding methods and executing resources should be specified. This live scenario serves as the background to identify all negative influences (*x*) in the process (**Process-Analysis**) on the problems (*Y*) (Example: xresource_01: Chef nibbles chocolate from the ingredients (90%)). Please be very specific here, because the results of the Input- and the Process-Mapping-Analysis are the rational base for the statistical tests later in the ANALYSIS phase.

But before we can analyze data, we need to collect data. And before we collect data, we should formulate hypotheses for the problems (*Y*) and their related influences (*x*).

The **Cause & Effect Matrix** (C&E Matrix) prepares these hypotheses. In the rows of the matrix all detected influences (*x*) from the input and the process are contrasted to the columns with all identified problems (*Y*). Experts now rate the degree of relationship between all influences (*x*) and all problems (*Y*). The result is a ranking of all influence–problem relationships ($Y = f(x)$), according to their strength (Example: xresource_01: Chef nibbles chocolate from the ingredients (90%) has a strong influence on (*Y*_01: Cookies taste bad; *must-be*; 20% fulfilled; CTQ).

At least it is your decision which of the evaluated *x*–*Y* relationships should be statistically investigated. In some cases, it might be easy to collect data from given databases or in other cases it might be difficult or even impossible to measure data. It might even be unnecessary to collect data at all, e.g., if you detected in your Process-Mapping-Analysis that there is no standard procedure defined to guide the activities, like recipes do. Simply by reasoning we know that varying personal preferences in performing the activities will lead to variations of the outputs and thus to problems.

Now the prioritized influences (*x*) and problems (*Y*) from the C&E Matrix are operationalized in the **Data-Collection-Plan**. This means that the corresponding measurands (e.g., time) and their units of measurement (e.g., days) are specified, as well as their scale level as: cardinal, ordinal, or nominal.

Based on this operationalization the statistical **Hypotheses** for *x* and *Y* can be formulated either as relationship-hypothesis: "There is a/no relationship between x and *Y*" or as a difference-hypothesis: "There is a/no difference in: *Y*s between: levels of *x*." The hypothesis and the scale level of x and Y determine the appropriate graphical representations and the appropriate statistical tests. If there are doubts whether the data will be reliable, especially if the data are gathered by different people with different expertise or ambivalent measurement standards, it is recommended to perform a measurement-system-analysis (MSA) to standardize the measurement.

Now, the data can be collected according to a sampling plan with a certain sample size, depending on the selected statistical test and a schedule, depending on the flow and properties of the data.

After the data collection, the MEASURE phase is completed. The **ANALYSIS** phase starts with a plausibility check of the collected data.

Depending on the scale level of variables, the evaluation of Min- and Max-values and the statistical parameters (mean, median, mode, standard deviation, range, percentiles) is recommended (**Data-Evaluation**). A graphical analysis of the data, with a check of their distribution (Histogram, Boxplot, Pareto-Chart) and their progress over time (Time-Series-Plot) complete these first inspections.

Now the **performance of the process** can be determined by calculating the process capability (Yield, DPMO, cp/cpk, Sigma Level) and by generating Control-Charts to determine patterns and signals in the data over time.

In the next step, the **Hypotheses** are statistically **tested**. Every statistical test gives you two important results. The p-value about the statistical significance lets you decide either to confirm ($p > 5\%$) or to reject ($p < 5\%$) the H0 hypothesis, in favor of the HA hypothesis. If you accept the HA hypothesis, the p-value tells you how reliable your statement about the tested difference or relationship is. But is your result not only statistically significant also practically relevant? This practical relevance concerns the size of the difference (e.g., difference in means) or the size of relationship (e.g., level of correlation). However, if you find a significant and relevant relationship between an influence (x) (e.g., the amount of chocolate) and an important attribute of the output (Y) (e.g., the taste of the cookie), then this might be the starting point for the search of underlying root causes.

In some cases, you might not have any data available for this deeper going analysis to explain causes (x') of the causes (x) of the problem (Y). A rational **Root-Cause-Analysis** can here support finding the root causes (x') of the influences (x) of the problem (Y). This is done in a cascade of "Why"-questions with the experts in your team, leading to a Hierarchy-Tree of causes (x) with the problem (Y) at the top and the root causes (x') at the bottom.

The root causes (x') are the anchor points for solutions in the beginning of the **IMPROVE** phase. A workshop with the experts might be sufficient to **develop solution ideas** to eliminate or circumvent the root causes (x') and thus to solve the problems (Y).

In those areas where technical parameters need to be adjusted, a **Design-of-Experiments** (DoE) might be helpful to systematically identify the best adjustments of a bundle of root causes (x') to optimize one or more parameters of the output (Y).

All solutions with a positive benefit/effort ratio, which are also supported by the Sponsor, are specified in an **Action-Plan**. Here you should be very specific about: *Who?—Does What?—Till When?* The level of concreteness of the action items often determines the success of their implementation.

There might, however, be failures or harmful secondary effects associated with the action items. The Failure-Mode-and-Effects-Analysis (**FMEA**) serves to reduce these risks.

After you have formulated appropriate countermeasures against the risks of your actions, you can implement both actions and countermeasures.

It is necessary that your Sponsor supports this implementation by informing and motivating the employees, especially those who were not involved in the project. At least it is important that the Sponsor publicly presents the planned actions and commits to control their implementation and their success.

After the implementation, new data for the new process are collected for the **CONTROL** phase. The process capability is again calculated (Yield, DPMO, cp/cpk, Sigma Level) and Control-Charts are generated to determine patterns and signals in the data over time.

With a **statistical test**, you should calculate the difference between the performance of the process **before and after its improvement**. Please find out whether this difference is statistically significant and practically relevant.

The degree of this before–after difference can help to calculate the **financial benefits** of the project. This is an important advantage of Six Sigma that you can reliably assess the actual results of the process and their effects on the company and on the customer.

To assure the sustainability of your improvements over time, a **Process-Management-Plan** should be developed. It supports to continuously monitor the process with Control-Charts and gives recommendations what to do, if the process gets out of control.

This completes the **Story-Book** of your project where you should continually record what you and your team did along the DMAIC phases. I hope that responsible managers and your Sponsor acknowledge your performance and the performance of the whole team. Maybe your Project-Story-Book becomes one deliverable of your Green Belt or Black Belt certification.

The project itself is finished here. And it is the start for a continuous improvement process, because no process is ever perfect. 6 sigma is a vision and a Six Sigma Project is just one step to this perfection.

Lean and Six Sigma: Partners for Quality 13

13.1 Lean and Six Sigma: Roots and Development[1]

Oh, you are already here. Sorry for wasting your time. And I am wasting my time. And wasting time is a sin in Calvinism. But what does Calvinism have to do with Lean-Thinking and Six Sigma? It was a trigger! This video is about the historical roots and the development of Lean-Thinking and Six Sigma.

The story begins in the sixteenth century with the Swiss-French theologian Jean Calvin (1509–1564). He was convinced that the fate of every human being is predestined by God even before his birth to belong either to the chosen ones or to the damned. The chosen ones receive salvation, the others the wrath of God for their sins (Fig. 13.1).

At that time, the individual economic success was interpreted as a visible sign of God's election. However, since no one could know whether he or she was chosen by God or condemned, and because people wanted to be publicly perceived as chosen, the diligence at work became a Calvinist principle. The followers of the doctrine worked very hard to earn money and thus to improve their public image. But not only the waste of time, but also every kind of luxury was considered a sin. Economic success on the one hand and the ban on luxury on the other led to a surplus of money. This money was reinvested in technologies, machines, and methods, which made the work more efficient, led to higher volumes and thus further increased the economic success. The diligence (Latin: industria) thus became a trigger of industrialization.

Machine power, the standardization of parts, and the division of labor belonged to the resources and methods for more efficiency. The disassembly lines in the slaughterhouses of Cincinnati and Chicago (USA) were the model for the assembly lines in automobile production, initially at Ford (1914) and Citroën (1919). Frederick Winslow Taylor's (1856–1915) Scientific Management had a strong impact on the division and sequencing of the work into only a few specialized operations for each

[1]Video 42 (https://www.youtube.com/channel/UCShsy50l5dFsoSXU__hBKSA)

© Springer Nature Switzerland AG 2019
R. Hutwelker, *Six Sigma Green Belt Certification Project*, Management for Professionals, https://doi.org/10.1007/978-3-030-31915-1_13

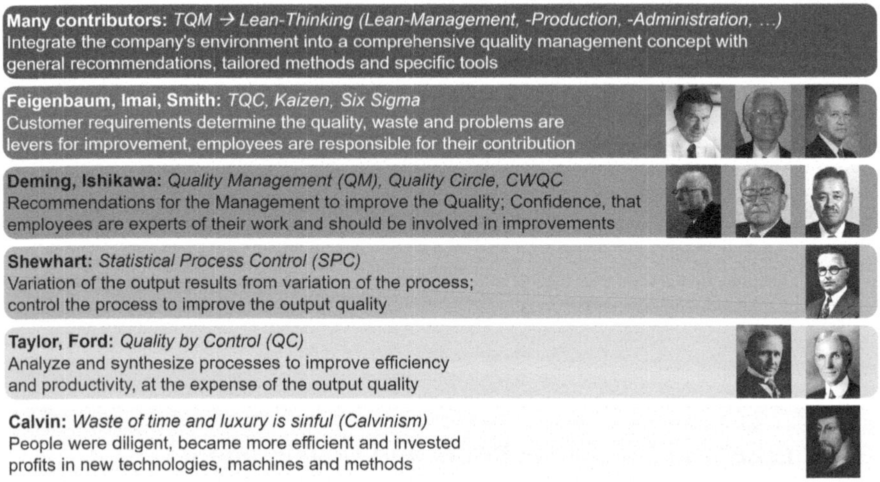

Fig. 13.1 Faith stimulated productivity and increased productivity required a management of quality (Picture references: Jean Calvin: https://commons.wikimedia.org/wiki/File:John_Calvin_-_Young.jpg, Frederick Winslow Taylor: https://commons.wikimedia.org/wiki/File:Frederick_Winslow_Taylor.JPG, Henry Ford: https://commons.wikimedia.org/wiki/File:Henry_ford_1919.jpg, Walter A. Shewhart: https://en.wikipedia.org/wiki/File:WalterShewhart.gif, W. Edwards Deming: https://commons.wikimedia.org/wiki/File:W._Edwards_Deming.jpg, Kaoru Ishikawa: https://commons.wikimedia.org/wiki/File:Kaoru_Ishikawa.jpg, Taiichi Ohno: https://en.wikipedia.org/wiki/File:Taiichi_Ohno.jpeg, Armand V. Feigenbaum: https://en.wikipedia.org/wiki/File:Armand_Vallin_Feigenbaum.gif, Masaaki Imai: https://commons.wikimedia.org/wiki/File:Masaaki_imai-1511438732.jpg, William B. Smith, Jr.: No source for copyright of the photo available)

worker, which had to be exactly repeated throughout the entire shift. According to Taylor, the work processes have to be subdivided to such an extent that even a trained monkey can perform them. Workers no longer had an overview of the subdivided process, no say, and no responsibility for the status of their work results. This had a negative effect on the quality of the final product. For this reason, Henry Ford (1863–1947) introduced a quality control system (QC) in which finished products were inspected and the defective products were sorted out and repaired. The movie: Modern Times (1936) by and with Charlie Chaplin (1889–1977) caricatures the consequences of the assembly line on the workers.

Quality control (QC) prevents the customer from receiving a defective product. But the reactive repair of a finished product is expensive. Walter A. Shewhart (1891–1967) recognized this, developed the Control-Chart and became a pioneer of statistical process control (SPC). The Control-Charts, installed at significant steps along the process chain, indicate the extent of conformity or deviation of the process from its specifications. Shewhart showed that reducing the variation of the process proactively led to an improvement of its outputs. And he implicitly founded the central principle of Six Sigma: $Y = f(x)$ with the idea to control the relationship between the influences in the process (x) on the properties of the output (Y).

William E. Deming (1900–1993) expanded the view from the quality defects of the products to the processes in which the defects were caused and from value-adding processes to support processes. He emphasized the role of management and formulated its tasks for quality management. Together with Kaoru Ishikawa (1915–1989), he was convinced that employees were the experts of their work and therefore had to be involved in the analysis and improvement of the respective processes. Now the entire company became the focus of improvements.

Ishikawa emphasized the importance of internal customer, developed the Fishbone Cause & Effect Diagram, which is an important tool of the Seven Tools of Quality. And he developed the idea of Quality Circles, in which employees systematically eliminate waste and contribute to the continuous improvement of their processes, according to the PDCA-cycle (Plan Do Check Act). The PDCA-cycle was developed by Deming/Shewhart and the definition of waste was introduced by Taiichi Ohno (1912–1990).

The PDCA-cycle is the predecessor of the DMAIC-cycle (Define, Measure, Analysis, Improve, Control) of Six Sigma. And the additional consideration of waste justifies the concept of Lean Six Sigma today.

Due to his expansions of Deming's concepts, Ishikawa is considered the founder of Company-wide quality control (CWQC), the Japanese way of Quality Management. In the USA Armand V. Feigenbaum (1922–2014) developed with his concept of Total Quality Control (TQC), a similar management system for the development, maintenance, and improvement of quality. In TQC each employee of a company is basically responsible for the quality of his or her work. This responsibility is regularly concretized in the quality-relevant objectives of the corporate units. Associated data are collected and checked in order to counteract deviations. The sources for the objectives are customer satisfaction and the company's quality costs.

In the following years, the American approach of TQC and the Japanese CWQC merged into the concept of Total Quality Management (TQM), which additionally considers the relevant environment of the company. Indicators for business excellence were developed and awards were given to the most competitive companies, like the Deming Prize in Japan, the EFQM Award in Europe, and the Malcolm Baldridge National Quality Award in the USA.

The Toyota Productions System (TPS), developed by Sakichi Toyoda, Kiichiro Toyoda and Taiichi Ohno, is currently considered the best practice of a Total Quality Management.

Masaaki Imai (1986) summarized the core of Japanese TQM in his book KAIZEN for a "change for the best." KAIZEN became popular in Europe and the USA as the basis of continuous improvement processes (CIP) and the core of "Lean" ideas. The term "Lean" encompasses various business areas. Lean-Management, Lean-Production, Lean-Administration but also, e.g., Lean-Development and Lean-Construction are commonly referred to as Lean-Thinking. And Lean-Thinking has recently replaced the term TQM.

Parallel to Masaaki Imai's KAIZEN, William B. Smith and Mikel J. Harry (1988) developed Six Sigma at Motorola. The trigger was a specific quality problem in Motorola's semiconductor division and one customer required a performance level of 6 sigma. This requirement is the origin of the developed method to achieve this performance level (personal message). The core of Six Sigma is the DMAIC-cycle, a problem-solving method with a chain of chronologically linked tools. For many companies, General Electric is the role model for the successful implementation of Six Sigma.

In summary, it can be said that faith in God stimulated productivity and that increased productivity required a management of quality. There are many different opinions on this topic and my brief overview thus also reflects my personal point of view. Nevertheless, it might help you to understand why both approaches are important today, why they have become necessary, and what their purposes are.

13.2 Lean-Thinking Vs. Six Sigma: Comparison of Attributes[2]

I have just taken a sample of job offers on the net for Lean-Thinking and Six Sigma. It shows that two thirds of the 30000 global vacancies last month (January 2018) were posted for Six Sigma and one third for Lean experts. Of course, this snapshot does not mean that one of the approaches is superior and that your company should decide for Six Sigma and against Lean-Thinking. It merely shows that companies currently have different needs. And these needs may have arisen from a comparison of the two approaches.

Although both Lean-Thinking and Six Sigma focus on customer satisfaction and cost reduction, there are differences in their properties. The following comparison of Six Sigma with Lean-Thinking (see Fig. 13.2) is based on their properties: Assumption, Objective & Potentials, Focus, Approach, Principles, Methods & Tools, Character and Advantages/Disadvantages. I will try to contrast both approaches, although their boundaries are blurred.

The approaches of Six Sigma and Lean-Thinking show differences in the roles and tasks. In Six Sigma, the employees, the management, and customer continuously identify problems in the products and services. Task of the management is to periodically prioritize and select appropriate problems for Six Sigma projects. Temporary project-teams, led by Green/Black Belts, solve the selected problems. In Lean-Thinking, the management is responsible to identify appropriate methods for the company. Experts then implement and adapt these methods. Employees in CIP-Teams continuously eliminate waste and standardize their process.

The principles of Six Sigma are to identify, collect, prioritize, and select suitable problems for projects, then to assign and—if necessary—qualify Green/Black Belts as project leaders, assign the subject matter experts for the project team and the Sponsor for each selected project. Then the project is implemented, and its financial benefit calculated. Lean-Thinking first identifies and prioritizes customer requirements (= value), then identifies and optimizes the value-stream as a chain of value adding process steps. The Just-in-Time rule is implemented by realizing the

[2]Video 43 (https://www.youtube.com/channel/UCShsy50l5dFsoSXU__hBKSA)

	Six Sigma	Lean-Thinking (Lean-Management, Lean-Production, Lean-Administration; ...)
Assumption	Unwanted variation of properties of the input (xi) and process (xP) leads to variation of properties of the output (Y) Reduce unwanted variation of input/ process to improve the output	Maximal plant utilization, push-principle and waste lead to a discontinuous value-stream Problems are opportunities and employees are the experts to solve problems in their processes
Objective & Potentials	**For selected products/ services:** **Reduce costs by:** Increase Quality and Availability, decrease Consumption **Increase customer satisfaction by:** Identification and fulfillment of customer requirements **Potential:** Costs of unfulfilled customer requirements and cost drivers	**For the entire company:** **Reduce costs:** Efficient (waste-free), stable, harmonized and standardized processes **Increase customer satisfaction by:** Achieve promised **Quality** & **Availability** of products/ services **Potential:** Costs of unfulfilled requirements, costs of cycle times, stocks, inventories and waste
Focus	**Entire organisation:** program to improve important (semi-finished) products/services **Project-team:** solve the identified problems of products/ services	**Entire organisation:** optimization of value-stream of processes in administration, production, service and support (...) **Work-unit:** continuous improvement in regular meeting employee teams (CIP-teams)
Approach	**Employees, Management & Customer:** identify problems in products/ services (continuous) **Management:** prioritize and select problems for Six Sigma projects (periodically) **Green/ Black Belt & Project-Team:** solve selected problem (temporary)	**Management:** identify appropriate methods for the company (once) **Experts:** implement and adapt methods (periodically) **Employees/ CIP-Teams:** eliminate waste, standardize process (continuously)
Principles	- identify, collect, prioritize and select suitable problems for projects - qualify Green-/ Black-Belts, members of project-team and a sponsor for selected project - implement the project - calculate financial benefit of project	- identify and prioritize customer requirements (:= value) - identify and optimize value-stream (chain of value-adding process-steps) - implement pull-principle and act order-related (JIT) - implement continuous-improvement-process (CIP, KAIZEN)
Methods & Tools	**Generic problem solving approach (DMAIC) with e.g.:** **Rational analyses of:** customer requirements, problem-cause relationships, root-causes **Statistical analyses of:** hypothesis on relationships (x-Y) and differences (Y1-Y2) **Optimization of specific parameters:** Design of Experiments (DoE) **Planning of processes:** process-simulation, DFSS	**Set of specific methods and recommended actions for a wide range of application:** **For Lean-Management:** Key-Performance-Indicator-System, Policy Deployment, Visible Management, Simultaneous Engineering, Decentral End-to-End Process Responsibility, Improvement-Teams, Incentives and Salary System **For Lean-Production, Service & Administration:** Value-Stream-Analysis and -Design, One-Piece-Flow, Kanban, Overall Equipment Effectiveness (OEE), Single Minute Exchange of Die (SMED), Total Productive Maintenance (TPM), 5S, Elimination of Waste and Standardization
Character	**Medical Therapy: Treatment for specific disease** 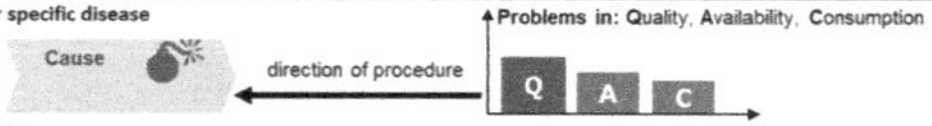 Cause direction of procedure Problems in: Quality, Availability, Consumption Q A C	**Fitness Program: Strengthening of body functions** 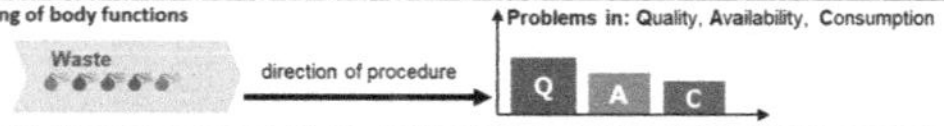 Waste direction of procedure Problems in: Quality, Availability, Consumption Q A C
Advantages & Disadvantages	+ financial benefits of projects calculable + identification of root causes for even complex problems - optimization of sub-processes without considering the overall context - the majority of employees are not involved in improvements	+ involvement of a large part of the employees - cultural change of company necessary + integrated optimization of all company divisions - decisions often based on perception and not on data

Fig. 13.2 Comparison of Six Sigma with Lean Thinking. Six Sigma goes deep, and Lean goes wide to reduce costs and increase customer satisfaction

pull-principle and the order-related orientation of all activities. CIP-Teams are established for the continuous improvement of their processes.

Six Sigma is characterized by its scientific methodology. The DMAIC is a generic problem-solving approach, with, e.g., rational analyses of customer requirements and the root-cause analysis. The focus is on statistical tests of hypotheses on relationships (x–Y) and differences ($Y1$–$Y2$). The optimization of specific parameters, e.g., of machines, is realized by Design of Experiments (DoE). Process planning is supported by process simulation and the Design for Six Sigma (DFSS) method.

Lean-Thinking uses pragmatic methods and sets of recommended actions for a wide range of application, for the management, for the production, for service and administration. Methods and tools for the management are, e.g., the: Key-Performance-Indicator-System, Policy Deployment, Visible/Visual Management, Simultaneous Engineering, Decentral End-to-End Process Responsibility, Improvement-Teams, Incentives and Salary System. Improvement of production, service and administration are based on the: Value-Stream-Analysis and -Design, One-Piece-Flow, Kanban, Overall Equipment Effectiveness (OEE), Single Minute Exchange of Die (SMED), Total Productive Maintenance (TPM), 5A/5S, Elimination of Waste and the Standardization of methods.

One can exaggeratedly say that Six Sigma has the character of a medical therapy: The illness of a person, as a specific deviation from the normal state—in the example a deviation in quality indicators of the output—becomes the starting point of the diagnosis of the influences and their root-causes, which leads to specific treatments of the involved organs and physiological processes. Lean-Thinking is more like a fitness program. All the organs and physiological processes that are important for certain required body functions are continuously strengthened and all influences that could weaken the body are eliminated. As a result, the output shows improvements in its indicators of quality, availability, and resource consumption.

Thus, the directions and purpose of both approaches differ: In Lean-Thinking from the elimination of waste in the processes to the improved outputs and in Six Sigma from the identification of a certain problem in the output to its specific root-causes in the process. This perspective polarizes the two approaches more than perhaps adequate. And the same applies to the last comparison of their advantages and disadvantages.

The advantages of Six Sigma are, e.g., that the financial benefits of projects can be calculated very easily in many cases and that root-causes of even complex problems can be identified. However, it can be criticized that projects sometimes only optimize certain parts of processes, without considering the overall context. And the projects overall only involve a small portion of the employees so that their capabilities for improvement remain untapped. In contrast, Lean-Thinking involves a large part of the employees. The introduction of Lean-Thinking, however, often contradicts the habits of employees so that a cultural change of the company becomes necessary. The methods allow an integrated optimization of all company divisions. Decisions are, however, often based on perception rather than on data.

Although both Lean-Thinking and Six Sigma focus on customer satisfaction and cost reduction, there are differences in their properties. However, in recent years Six Sigma has been combined with Lean-Thinking, leading to Lean Six Sigma. On the other hand, Six Sigma can meaningfully complement any company's Lean-Thinking system. (Lean-Thinking and Six Sigma are therefore not opposite, but complementary approaches.)

The implementation and sustainable maintenance of Six Sigma and Lean-Thinking in a company encounters a variety of resistances, requires a lot of effort and their benefits are only gradually becoming apparent. Persistence of the management is one success factor. And at a congress in Amsterdam, Jack Welch answered the question of how he managed to make Six Sigma so successful at General Electric by saying: "You have to reward it." The successful implementation, however, offers an advantage that cannot be easily copied by competitors. Each company has to find its own way to do this.

References

Harry, M. J. (1988). *The nature of six sigma quality*. Rolling Meadows: Motorola University Press.
Imai, M. (1986). *Kaizen: The key to Japan's competitive success*. New York: McGraw-Hill.

14 Tips for Six Sigma Sponsors to Support Projects and the Program

Dear Sponsor,

Probably and hopefully you are responsible for the process and its outputs for which your Green Belt is to carry out a Six Sigma project. You do not need to master the Six Sigma methods, but you should support the project with your influence as a manager. Your role in a Six Sigma project is as important as your Belts. These tips can support you to take on this role optimally.

In these tips I mainly refer to Green Belts, but they are also mostly valid for Black Belts. However, I recommend supporting the certification of your Green Belt first and not to start with the Black Belt right away. Here you can first find out if the candidate has the competence and develops a passion for Six Sigma. Only in a second project the development to the Black Belt should take place, without the risk of a wrong investment of time, money, and frustration.

If you are interested in the certification project of your Green Belt, read the tips 1–3. The tips 4–9 might be helpful, if you want more.[1]

14.1 Six Sigma for Sponsors and Basic Recommendations

Six Sigma is a method for targeted improvement of repetitive processes in production and service. It offers a kit of chronologically linked rational and statistical tools for problem-solving to increase customer satisfaction and to decrease costs.—What do we understand by a process?

[1] The tips are partly based on the results of S. Schindewolf's unpublished Master's thesis (2011) on Six Sigma programs in 78 German companies. Selected results of this thesis were published by Leyendecker et al. (2011). The tips are also based on new, unpublished results, based on the data collected by Schindewolf.

© Springer Nature Switzerland AG 2019

R. Hutwelker, *Six Sigma Green Belt Certification Project*, Management for Professionals, https://doi.org/10.1007/978-3-030-31915-1_14

Each value of a company is created in a process. An input (e.g., an order/ingredients) is transformed into an output (e.g., a cookie) by activities. The activities are guided by methods (e.g., a recipe/hygiene standards) and executed by resources (e.g., a chef/scale/oven). Every processes is based on this structure.—What problems can be solved?

Six Sigma projects start with a specific problem of the output. If the cookies taste bad, have too many calories or too much weight (quality problem), are delivered too late, the quantity is not sufficient (availability problem) or the energy consumption and the working hours are too high (consumption problem), then these problems (Y) provide anchor points for a Six Sigma project.—What is your cookie?

Every form we fill out, every answer we give, every decision we make, and every preliminary product we create is an output. If at least 30 outputs are created per month, then the process is suitable for Six Sigma. Tooth brushing would also be suitable. The best practice has even been defined for this process in order to achieve the optimum output. And everyone knows the problems and costs that can result from a bad brushing process.—How to implement a Six Sigma project?

Process experts first identify and evaluate the triggering influences of the bad tasting cookies (Y) in the methods, resources, and inputs (x). Then these influences (x) and problems (Y) are measured for about 1 month. Next the data are statistically evaluated, and the root causes (x') of the triggering influences are identified. Solutions are implemented to eliminate, circumvent, or adjust the root causes (x'). Now, data are measured again to assess the success and prove the financial and other benefits. Finally, the process owner receives control charts to monitor the process and a reaction plan for new problems to ensure sustainability.—Do you have process responsibility?

Process owners are potential Sponsors of a Six Sigma Green Belt project. They assess the success of each project phase and decide on the implementation of measures. And Master Black Belts supervise these projects, from the identification of the topic to the verification of its success.—What can you expect?

The financial potential of our Green Belt certification projects vary between $10,000 and $500,000. (See list of project topics.) The benefit of our certification projects thus clearly exceeds the course fee and the indirect costs of the project.—What should you keep in mind?

Eight Basic Recommendations for the Management of Six Sigma

1. **Challenges and problems**

   Managers do not like problems, but they love challenges. From a Six Sigma perspective, it makes no difference whether you sometimes brush the bar today in terms of quality, or whether you want to raise the bar to provide more outputs tomorrow with the resources available. Six Sigma analyzes the determinants of today's performance and seeks solutions to improve it.

2. **Sources for Six Sigma projects**
   In the beginning of Six Sigma it is sometimes difficult to identify suitable topics for projects. But there are four sources: Management (performance indicators and target achievements), Customer (ratings and complaints), Employees (observations and clarifications), and Systems (suggestion system and results from value stream analyses). Filter this information according to their relevance to important processes/outputs, assess their potential and prioritize them. These can be your topics for Six Sigma.
3. **Improve process chains upstream**
   Focus on complete process chains, from the last output that leaves the company to the first input that the company receives. Split the process chain into functional units and start the first project with the last output. This is where all deviations in quality, availability, and consumption accumulate. Solve these problems as far as possible and start the next project with the deviations of its input, which is the output of the upstream process step. Work your way upstream to the first external inputs and integrate the supplier of this output into the last project of this process chain.
4. **Monitor process performance**
   The sigma level is a dimensionless measure to determine the process performance. At project start we often measure negative sigma levels. And 6 sigma is a vision, rarely a target for the next year. If, however, all (intermediate) outputs of a process chain are evaluated according to the same performance measure and monitored by comparable control charts, then the probability of cooperation across departments increases because everyone is pulling in the direction of the common performance indicator. And this could be the sigma performance indicator.
5. **Standardize the methods**
   Many companies only focus and document the activities of their processes. Methods that guide these activities are rarely mentioned. And practice shows that many activities are carried out according to personal expertise and preferences. This is understandable in innovative and creative processes, but not in standard processes, if, e.g., different chefs bake their cookies without a recipe. It is the methods that bring intelligence and hopefully even best practice into the process.
6. **Unmask solutions that are disguised as causes**
   Only causes of problems are suitable interfaces for solutions. How do you react when an employee tells you: "I came to work too late today because I don't have a

Ferrari"? – This explanation is of course nonsense, because the missing Ferrari is not a cause but a solution that does not (yet) exist. Please pay attention to how often the occurrence of problems is explained by desired solutions. You can identify them by negations containing the prefix: un- or the words: no or not. In my experience, there are many Ferraris in companies although no one has ever observed how a non-existent Ferrari led to a delay.

7. **Acknowledge success**

   Jack Welch—a former CEO of General Electric—once answered the question of how he managed to make Six Sigma so successful in his company by saying: "You have to reward it." With successful programs, each Green Belt carries out an average of 1.6 projects per year and the efficiency of implementation increases with experience. For less successful programs, each Green Belt only carries out a total of 1.2 projects during the life span of the Six Sigma program.

8. **General fitness and specific treatments**

   Six Sigma and Lean-Thinking do not compete but complement each other. Lean-Thinking is suitable for increasing the fitness of processes. Improvements of the process are based on concrete, scientifically proven recommendations: Do this and leave that. This widely increases the quality and availability of the outputs and reduces their consumption of inputs and resources. Six Sigma is the treatment for specific problems of the outputs. It develops tailored solutions for their causes in the process. Both methods are the part of our program: Six Sigma and Lean: Quantitative Tools for Quality and Productivity on https://www.edx.org.

From the perspective of Six Sigma, it does not matter whether the Project-Topic is a cookie, a component, a list, or just a data entry (Sect. 4.3). Each is an output that more or less meets the requirements for quality, availability, and consumption. And each of these outputs is preceded by activities and inputs that positively and negatively influence the outputs. This is what Six Sigma is all about.

The following list of selected Green Belt certification projects provides an insight into the application range of Six Sigma (Table 14.1).

Table 14.1 List of selected Green Belt certification projects, demonstrating the application range of Six Sigma

Agriculture:	Optimise sowing and harvesting cycles in a tulip farm
	Rubber losses during harvesting
Chemistry/ Drugs:	Optimize pump maintenance intervals
	Reduce effort/ errors in recipe-specific drug production
	Reduce the variation of the colours of car coatings
Construction:	Accelerate service contracts to establish department stores
	Quality defects and delays in delivery of precast concrete parts
Finance:	Cycle times and compliance with rules for credit decisions
Information:	Errors in translation and legal information of multilingual webpages
	Increase permission rate/ channels for customer contact
	Optimize bonus distribution for customer types
	Abandonment rate in online payment process
	Reduce contact frequency of frequent callers in hotline
	Reduce fraud rate on game/ video downloads
	Reduce fraud rate on prepaid contracts
	Reduce waiting times in hotline queue
Manufacturing:	Bristle losses in street cleaning machines
	Improve maintenance orders for helicopters
	Contact faults in wiring harnesses for cars
	Reducing waste of cutting insulating blankets for aircrafts
	Decreasing performance of condensate removal pumps
	Delayed drawing creation for chimney pipes
	High reject rate of gravity die cast pistons
	Incident management errors and delays for automotive parts
	Increase start-up speed in paper production
	Length deviations for aluminium rods
	Material losses during vulcanization of tyres
	Pinholes in blister packages
	Reduce scrap rate for injection molded parts
	Increase reusability of masks in semiconductor manufacturing
	Scratches and mounting errors on bicycles
	Sterilization of medical products
	Errors and high expenditure of time with storage of supplied parts
Mining/ Oil/ Gas:	Problems resulting from drawing standards of deep sea supply units
	Improve filling of LPG cylinders
	Reduce errors in welding seams
	Scope of component maintenance of oil platforms
Retail:	Secure increasing order volume with constant resources of car sales
Technical Service:	Increase OLED lifetime
	Resource optimization in prototype development

14.2 Focus and Effects of Projects

Most Six Sigma projects focus on **costs and revenue losses** that will **result from quality defects** and **from time losses**. Some projects focus on **costs for resources** that should be **kept constant despite the expected increase in order intake/ volume.** This applies to **production and services**.

Sometimes there are misunderstandings about the scope of a project and managers wonder why the hypotheses in the MEASURE and ANALYSE phase do not directly relate to costs. And in this context, they sometimes refer to the magic triangle. The term: Magic Triangle (Triple Constraint, Iron Triangle) is used both in project management and in production. It represents the three dependent variables: *Quality*, *costs*, and *time* and their relationship in a project or production plan. It is assumed that improving the quality of a product will lead to an increase in time and costs. Improving availability will reduce quality and increase costs. And the reduction of costs will probably reduce the quality and increase time. The challenge is to align and balance the objectives for these variables.

In Six Sigma a different perspective makes more sense. Because from the perspective of the Process-Problem-Solving Model (Sects. 2.4.2 and 2.4.3), the costs are indirect financial indicators at the management level, derived from the direct physical indicators of quality, availability, and consumption at the operational level (Fig. 14.1).

Each Six Sigma project starts with a problem, i.e., the deviation of a (intermediate) product or service in at least one important attribute. According to the attribute there are three types of problems: Deviations in the quality and in the availability of the output and the excessive consumption of inputs and resources within the process. The core of a project is therefore about these topics:

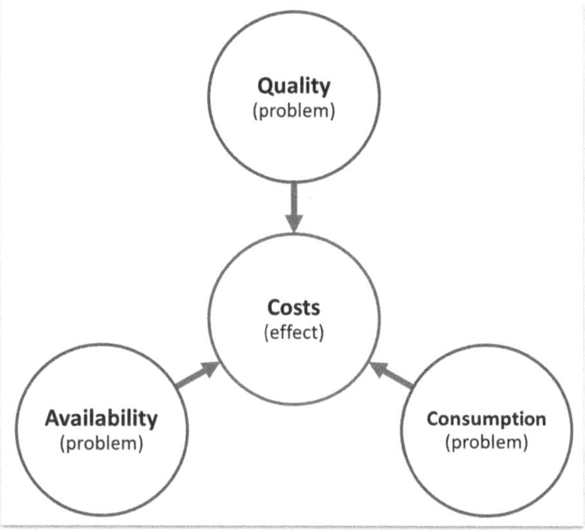

Fig. 14.1 Relationship between problems in the quality, availability, and consumption of products/ services and their financial effects

Quality: Output is wrong, defect, bad or faulty, not or only partially suitable for the targeted purpose (e.g., cookies taste bad).

Availability: Output is not available in the right quantity (too much/too few), not available at the right time (too early/too late) (e.g., cookies are delivered too late).

Consumption: Activities to generate the output are time consuming, there are waiting times in between or they are wasteful in consumption of inputs or resources (e.g., ingredients and energy are wasted, too many chefs for too few cookies).

The **effects of these problems** are **dissatisfied (internal) customers** and **additional costs** for the company.

In the beginning of the DEFINE phase, the suitability of the identified problems is checked and their effects on costs and customer satisfaction is estimated (Sects. 4.2 and 4.3). The targets of the Project-Charter always reflect these problems, upside down (Sect. 6.3). A Six Sigma Belt improves the process (baking cookies) to reduce the problems (Cookie taste bad) but is not responsible for the resulting cost reduction. Of course, an improved process, leading to excellent cookies, will contribute to the bakery's monetary profit and customer satisfaction. But no Green Belt should accept to analyze the relationship between the indirect measure of the cost of cookies and the direct causes of problems in the process. Prediction accuracy would be wasted here. Therefore, Six Sigma links problems of the outputs with the causes in the process and not with the costs resulting from the problems. Please make sure that problems and their effects are not mixed up.

At the end of the CONTROL phase, the financial benefits are calculated, and additional benefits are determined. A Belt is expected to prepare a proposal for the calculation of financial benefits (Sect. 10.4), preferably with the assistance of the Financial Controller. The benefit of the project may be less than the estimated potential. But at least the direct and indirect costs of the project should be considerably lower than the benefit from the project.

The **Green Belt is responsible** for **solving the problems** and **achieving the targets**. And the **Sponsor is responsible** for the **estimation of the financial potential** and the **achieved financial benefits**.

14.3 Role and Tasks for Sponsors

As a Sponsor, you should be responsible for the output problems that are in the project focus, as well as for the important triggering causes in the upstream process. This is the best prerequisite to improve the process and to win over the people involved. In my opinion, the term Sponsor is somewhat misleading. Typically, a Sponsor pays for access to commercial channels to advertise for himself. For me the terms mentor and promoter fit better to the role and tasks in the project.

The Green Belt is responsible for the methodical correct implementation of the project. The **Sponsor**:

- **Decides** on the targets, scope, adjustments, and implementation of measures
- **Supports** the cooperation with experts, establishes contacts, e.g., to (internal) customers and (internal) suppliers of the process and promotes the implementation of measures
- **Monitors** the progress of implementation and the sustainability of success.

This cooperation of a method promoter (Green Belt) and a power promoter (Sponsor) is a success factor in every project. However, I have met many Sponsors who reluctantly participate in the DMAIC phase steering and support the implementation little or not at all, although they have decided for the measures. This is hardly understandable because in the end they are responsible for the process and the success of the project.

The following table, also included in sigmaGuide, shows the typical checks of a Sponsor in the DMAIC phases. You can use it to **evaluate the progress of the project and the current scope**, e.g., if your Green Belt presents the results, documented in the Project-Story-Book, at the end of each phase. Here you should **decide on the success and the continuation with the next DMAIC phase and adjust the scope if necessary or recommend corrections and reworking for the current phase**. A Master Black Belt should have previously assessed the methodological basis of the results achieved (Fig. 14.2).

In addition to the checks let me summarize seven important tasks for you:

1. Decide on the implementation of the defined project (DEFINE: Project-Definition).
2. Permit and establish necessary contacts for interviews with (internal) customers (DEFINE: Voice of Customer).
3. Make sure that the scope and the associated targets in the project charter are realistically achievable and therefore motivating.
4. Support the two workshops (MEASURE: Process-Mapping/-Analysis; ANALYSE: Root-Cause-Analysis):
   (a) Make sure that the necessary experts are invited.
   (b) Indicate the importance of the project in the Workshop kick-offs and
   (c) Evaluate the presented results at the end of the workshops.
   (d) Your continuous participation in the workshops is not necessary and not recommended.
5. Decide on the implementation of the proposed measures (IMPROVE: Action-Plan).
6. Inform, motivate, and empower affected employees before implementing the measures (IMPROVE: Action-Plan).
7. Monitor the process performance, e.g., via the corresponding control charts, visualized on a wall of your office (CONTROL: Process-Management-Plan).

Six Sigma gives you access to the best method that process management has to offer, and most important: You are in control. Give your Green Belt the freedom to lead the project methodically and support him/her in overcoming cultural and

Phase	Tool	Tasks	Tips	Your check
Define	Part 1	**Identify a topic for a Six Sigma project**		
	Project-Topic	Decide whether the Project-Topic should be examined in more detail (Project-Definition).		yes
	Part 2	**Define a Six Sigma project**		
	Project-Definition	Decide whether the defined project should be carried out as a Six Sigma project.	Candidate should be motivated and like to work methodically and with numbers.	yes
Define	Part 3	**Implement a Six Sigma project**		
	SIPOC	Check whether the SIPOC between the first input and the last output includes all critical triggers of the problem.	If the critical influences already lie in the first input, the project field should be expanded or shifted upstream.	yes
	Voice to Critical	Check whether the Belt really interviewed all affected (internal) customers and managers.	If necessary, make contact with the interviewees and open the doors for the Belt.	yes
	Project-Charter	Decide on the selection of problems, the focus of the project, its targets, the team composition and the necessary budget.	To emphasize the importance of the topic, the Project-Charter can be signed by all involved stakeholders.	yes
	Stakeholder-Communication	Check whether you will be informed regularly and appropriately about the project.		yes
	Phase-Steering-Meeting	Decide whether you are satisfied with the results of the current phase and whether you want to start the next phase.	It would be normal if something had to be reworked or if a certain aspect had to be emphasized in the next phase.	Go to the next Phase
Measure	Input-Analysis	Check if there are critical inputs outside your responsibility.	Establish the contact to the (internal) supplier of the input and win him/her for the project.	yes
	Process-Mapping and -Analysis	Check in advance if the workshop is adequately prepared and all necessary experts are invited.	If possible: Open the workshop, emphasize its meaning and let the results be presented to you at its end.	partially
	C&E Matrix xY	Check whether in your opinion all important relationships between influences and problems have been established.	Have the C&E Heatmap explained to you and the summary of the C&E Matrix.	yes
	Data-Collection-Plan	If you are interested, then let the Data-Collection-Plan and the evaluation strategy be explained to you.	Data collection can lead to resistance and legal conflicts. You may have to support the collection of the data.	yes
	Hypotheses	If you are interested, let the hypotheses and the statistical tests be explained to you.		partially
	Phase-Steering-Meeting	Decide whether you are satisfied with the results of the current phase and whether you want to start the next phase.	It would be normal if something had to be reworked or if a certain aspect had to be emphasized in the next phase.	Rework the current Phase
Analyse	Data Evaluation	Check whether all relevant data were collected and the plausibility of their graphical representations.		...?
	Process Performance	Check process capability (Sigma-Level) and process performance over time (Control-Charts).	Some processes show a negative Sigma-Level in this phase. 6 Sigma is a vision and probably not the target of this project.	...?
	Hypotheses Tests	Check whether the most important hypotheses (C&E Matrix) have been statistically tested, interpreted and concluded.	Pay attention to the practical relevance (strength of a relationship or size of a difference) of significant results.	...?
	Root-Cause-Analysis	If possible: Open the workshop, emphasize its meaning and let the results be presented to you at its end.	Check whether in your opinion all important relationships between influences and problems have been established.	...?
	Phase-Steering-Meeting	Decide whether you are satisfied with the results of the current phase and whether you want to start the next phase.	It would be normal if something had to be reworked or if a certain aspect had to be emphasized in the next phase.	...?
Improve	Solutions	Check if the solutions are capable of eliminating, circumventing or adjusting the Root-Causes.	Pay attention to possible resistances against the solutions (complexity, costs, people).	...?
	Action-Plan	Check whether the measures fit the solutions, are sufficiently concrete and fairly distributed (who, what, until when; costs).	As a Sponsor, you decide which measure is (not) implemented.	...?
	FMEA	Check whether the most important risks of the measures have been evaluated and countermeasures formulated.	The countermeasures should be part of the Action-Plan and also formulated in concrete terms (Who, What, until When; Costs).	...?
	Implementation	Check regularly the progress of the implementation of the measures on the basis of the Action-Plan.	In order to reduce resistance, inform the persons affected about the measures, motivate and qualify them if necessary.	...?
	Phase-Steering-Meeting	Decide whether you are satisfied with the results of the current phase and whether you want to start the next phase.	It would be normal if something had to be reworked or if a certain aspect had to be emphasized in the next phase.	...?
Control	Data Evaluation	Check whether all relevant data were collected and the plausibility of their graphical representations.		...?
	Process Performance	Check process capability (Sigma-Level) and process performance over time (Control-Charts).	Some processes show a negative Sigma-Level in this phase. 6 Sigma is a vision and probably not the target of this project.	...?
	Hypotheses Tests	Check whether the most important before-after differences were statistically checked, interpreted and conclusions drawn.	Pay attention to the practical relevance (size of the difference) of significant results.	...?
	Process-Management-Plan	Check whether a Control-Chart and its maintenance is defined for each critical measurement type (Who, What, How often).	Ensure that the Control-Charts are visible and understood by those affected. Stick the control charts in your office, too.	...?
	Summary and Benefits	Check the plausibility of the representation (problem - cause - solution - benefit) and the amount of benefit together with your financial controller.	Acknowledge the success of all involved. How about a bonus?	...?
	Phase-Steering-Meeting	Decide whether you are satisfied with the results of the current phase and whether you want to start the next phase.	It would be normal if something had to be reworked or if a certain aspect had to be emphasized in the next phase.	...?
General	Lessons Learned	Check whether, in addition to lessons learned, topics for future projects have also been identified.	If you liked the project, you can order the next one right away. Maybe you can think of a competent Belt as a project manager.	...?

Fig. 14.2 Checklist for the Sponsor to monitor the progress in DMAIC phase steering presentations

organizational barriers. If you expect an excellent performance, then you should behave exemplary.

14.4 Selection of a Green Belt Candidate

Some companies consider the Six Sigma certifications and the role of a Belt as a career step for selected employees. This makes sense because it is not a disadvantage for a future manager to know methods to analyze and improve processes, which are the veins of value creation in every company. These criteria can support the selection:

1. Ability to objectively **observe**, **describe,** and **analyze** situations and developments
2. Interest in **dealing with numbers**, e.g., already working with Excel functions
3. Ability to lead a **team** with a balance between **perseverance** and **adaptability.**

Leyendecker et al. (2011) proved the superiority of aptitude-oriented selection. Green and Black Belts, which were selected according to certain criteria, realized about twice as many projects and the share of their successful projects was about 15% higher, compared to Belts of those companies which did not apply a specific selection.

Expertise on the topic itself is not required but can be helpful in the certification project. In the course of the project the Green Belt candidate acquires the competence to transfer the methods to topics outside their field of expertise. Thus they take the role of an implementation-oriented internal consultant.

And this is what this course is about:

- getting to **know** and **applying** the chronology of the basic Six Sigma tools and
- **transferring** this structure and sequence to similar projects and thus acquiring competence.

Six Sigma then unfolds its strength in the company if the Green Belt implements further projects, even outside his/her field of expertise (#9, #10).

14.5 Human Factors: Round or Square Wheels for Improvements

Many people strive to become better and better, in everyday life and at work. They develop habits and refine them bit by bit into automatized routines. The excellence of these routines corresponds to the ability to recognize and master requirements, of personal importance of these requirements and sometimes of external support. In order to improve we must realize that there is an undesirable deviation, a problem

Fig. 14.3 Human factors support and prevent change

What we can do quite well.

		Detect ...	Solve ...
... issues with ...	... others	✓	
	... ourselves		✗

What we can't do so well.

when we jump under the bar or a challenge when we want to raise the bar. We must be able and willing, and we should be allowed to solve it.

The more we have developed a habit, the more difficult it is for us to detect and accept resulting issues, and the more difficult it is for us to find solutions. This marks the limits of our ability, motivation, and legitimacy. And these limits are touched by all people in their working lives—some are already there, others are still far away (Fig. 14.3).

On the other hand, it is often easy for us to detect the issues of others. We only need to compare them with our strengths. But we should not expect gratitude when we point out their issues or even suggest solutions. Most people like to judge for themselves whether they have a problem at all and whether they want to solve it. And a suggestion from others can lead to shame about the problem, to fear of not coping with it, and anger about the interference.

An example of Lean-Thinking in the office: In my online coaching I often see a candidate executing even simple software commands—e.g., *Copy & Paste*—via the menu with the mouse. Waste of time, because the corresponding shortcuts are much faster. What happens when I point this out?

Grateful acceptance in rare cases: "Oh, thanks for the tip, I didn't know there were shortcuts". More often there will be rejections, like: "I know, but I rarely need to *Copy & Paste,*" "I prefer the commands via the menu and there I have all options at a glance," "I do it this way and that way," "The time saved with the shortcut is so small that it's not worth mentioning." My suggestion will be trivialized or rationalized away. If I start a discussion here in order to be right, then I will poison the climate very quickly: "Take care of your own work." By pointing out problems we are touching on a very sensitive spot in other people. Are we different ourselves? How would we react if someone attributed a problem to us and pushed for a solution?

There are also managers who do not like problems: "With us there are no problems, only challenges." In this way, a necessary duty is re-labelled as a voluntary exercise, because problems must be solved, challenges can be rejected. I do not want to condemn this tendency to trivialize through linguistic acrobatics. The term *challenge* sounds more diplomatic than *problem* and leaves us with a face. But managers act as role models and thus determine the attitude of employees. And for many, the statement "Houston, we have a challenge" is not very motivating. In many processes, however, fierce competition leads to a continuous struggle for qualification rather than for the gold medal. And this is a *problem* that must be solved and not a *challenge* that can be voluntarily taken.

But even a positive attitude toward problems is not sufficient to initiate improvements. A few years ago, a COO announced that the company still knew too little about customer requirements. And he asked employees to analyze the company's big data pool. Nothing happened, because the employees did not know exactly *what* to do and *how* to do it. And they pointed to the experts, whose job it was.

For better orientation, another company gave out a booklet to its employees with examples of typical problems in different processes and asked them to look for this in the working environment. At the same time, the improvement suggestion system was expanded for problems for which there is no solution yet. All employees now knew more precisely what problems the company was interested in and could detect them more easily. These problems are regularly evaluated and addressed as topics by Continuous Improvement Teams or Six Sigma projects. This strategy created a pull that is more motivating than a push.

But even that does not guarantee the desired improvements. The selected topics are handled in a working environment where problems are played down, habits defended, and initiatives mocked. Some employees are amused: "Kindergarten," others annoyed: "We don't need that, because we all improve our processes every day, without any request or guidance" and experts defend their advantage in knowledge and hide information.

A Green Belt should win the support of its environment and prevail against this resistance (Sect. 6.4). Your support could be to:

(a) Tell affected employees what the problem and its effect is about (**inform**)
(b) Provide the skills to successfully manage the changed demands (**qualify**)
(c) Demonstrate the individual benefits of the solutions (**motivate**) and
(d) Support people to assert themselves against rumors and resistance (**authorize**).

Inform, qualify, motivate, and authorize - in this order, an almost trivial insight but a noble task for a Sponsor.

14.6 Sources for Project-Topics in Your Company

Six Sigma, like Lean-Thinking, is implemented by many companies as a comprehensive improvement program (Sect. 13.2). This program can be designed as a mechanism with a continuous stream of new improvement topics (Sect. 2.2), which are evaluated according to their suitability and potential (Sects. 4.2 and 4.3) and then implemented as projects. Four **sources** feed this stream: **Management, customer, employees, and specific analyses**.

Originally, only strategically relevant topics were to be suggested that had to be accepted by top management. This severely **restricted the subject area** and some of these **projects were too complex**, others **unsuitable for Six Sigma**. Nevertheless, such topics were sometimes accepted by the experts, in order not to contradict the management and to point out the power of Six Sigma, and they failed. Consequently some companies were disappointed with the benefits of Six Sigma. For others, the stream of new project topics dried up and the program was terminated.

Leyendecker et al. (2011) investigated the impact of different sources for Six Sigma Project-Topics. They showed that the management with their strategic topics is a significantly more important source for Six Sigma projects than customer complaints, employee observations, and suggestions from the improvement system (Fig. 14.4).

I have never accepted focusing on strategically relevant issues. It ignores that each activity of a company is part of a process and that a chain is only as strong as its weakest link. That leads me to these questions:

- **Which of your processes are strategically relevant?**
- **Which processes are necessary for your company?**
- **Do you know processes in your company that are unnecessary?**

Cookies are strategically relevant for a bakery. And their taste has strategic relevance. But cookies of the best quality are useless if the order acceptance, sale, or delivery does not work. Such topics are sometimes addressed by Continuous Improvement Teams, although they exceed their problem-solving competence. And the value stream analysis also sometimes identifies topics that are predestined for Six Sigma.

The improvement suggestion system is unsuitable in its original design as a source for Six Sigma topics—understandable, because this is not its primary purpose. The system addresses employees and collects their solution ideas. **But a Six Sigma project always starts with a problem for which there is no solution yet**. However, filtering employee suggestions can identify appropriate and valuable problems.

Some companies even go one step further and opened their suggestion system to problem reports. They were dissatisfied with the acceptance, processing, implementation, and with the benefits of submitted suggestions because exclusive recognition

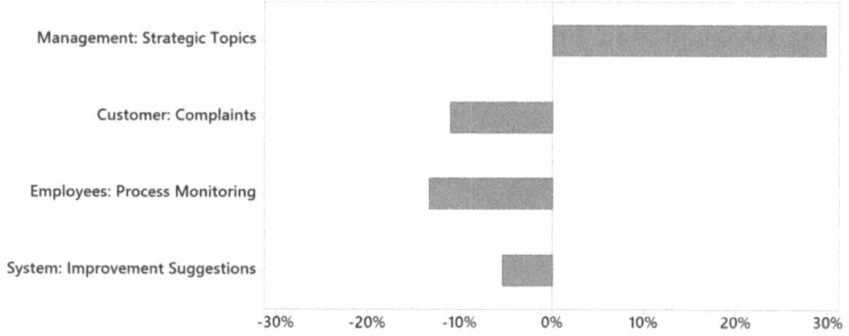

Fig. 14.4 Relative importance of different sources for Six Sigma Project-Topics in German companies ($N = 78$). The management with their strategic topics is the most important source ($p = 0.5$). The difference between the other sources is not significant (Multiple choice; Chi-Square Goodness-of-Fit Test, new results based on Schindewolf's (2011) data)

of solutions presupposes that the employee has first identified a problem and then analyzed its causes. Thus, the suggestion system is limited to ideas for simple problems. **Opening the suggestion system also for problems with unknown causes and solutions uncovers topics with a higher potential.** The incentive for employees is that they should solve selected problems themselves as part of Six Sigma training and receive a share of the benefits in addition to their certification.

I would **use any source** of project topics to get an overview of the potentials and **select the most promising ones** after evaluating them. Only when I know my deficits can I develop in a targeted way.

14.7 Sources for Project-Topics in Your Area of Responsibility

Of course, you do not have to involve the entire organization right from the start (*#7*). It would be enough to find one suitable topic in the beginning. However, this is also not always that easy.

A guiding principle in Continuous Improvement is: *Problems are opportunities*. Some managers, however, believe that a (still) unsolved problem is an indicator of their inability. Thus they are reluctant to admit problems or offer a whitewashing perspective on their processes. I very often experience this in Six Sigma Sponsor workshops. Here I always start with two questions.

1. **"How satisfied are your (internal) customers with the results of your work?"**
   The answers from the majority are typically: "above average" and "very" satisfied.—First, a contradiction is revealed in these results, because the majority cannot deliver above-average performance. But that is not surprising, because in front of colleagues, no one likes to admit problems. However, the reasons given by managers for their assessments are interesting: A small percentage of managers refer to regular surveys of internal customers and another small percentage to fulfilled service level agreements. The most common response is: "My customers don't complain."

With the second question I jump from the output to the input, which is of course the output of the next upstream process step.

2. **"Which portion of your working time do you typically spend ensuring the prerequisites that your processes require?"**
   The answers vary between 40% and 60% for managers and 15–25% for Green Belt candidates. My follow-up question, whether the necessary clarifications are systematically investigated, is usually denied.

What Do the Answers to the Two Questions Together Mean?
We usually only receive a complaint if an attribute of an output (quality, availability, consumption) lies outside the specification. German schools, e.g., send out a letter of complaint with a reference to the endangered transfer in the case of critical grades.

And parents will typically not align the goals with their children at the specification limit of school grades because satisfaction or even great satisfaction is linked to a significantly better performance.

This is not different in everyday professional life. Thus I think it is risky to conclude from the absence of a complaint that customers are satisfied or very satisfied with our outputs.

If we, on the other hand, spend 15–25% or 40–60% of our time with our inputs to ensure necessary conditions for our processes, then we are not enthusiastic about it, but we are so used to it that we do not complain to our (internal) suppliers. And they interpret the satisfaction with their performance as above average.

In other words: All respondents have a great need to clarify their inputs but assume that their outputs lead to high satisfaction. In a process, however, everyone has both roles as supplier and customer. That is why it is likely that our outputs, of which we are proud because we do not know otherwise, need to be clarified by our customers who receive them as inputs.

In this way, there is potential for improvement moving through the process chain, from the first input to the last output, which might be worth systematizing and prioritizing. Some of these problems and their causes are opaque and cross-linked and cannot be solved by quick fixes. In these cases a Six Sigma project can be useful.

If you first went to focus on your area of responsibility, then start with the last output and then go upstream to the inputs entering your process from outside. This may be the field for a suitable Six Sigma project. A Master Black Belt (we) can then review the topic and ensure in advance that this project does not become too complex for the certification and that on the other hand the benefits clearly exceed the course fee and further related expenses of the company.

14.8 Factors of Successful Six Sigma Projects and Programs

As a Sponsor, the success of the certification project of your Green Belts will probably influence your attitude to Six Sigma. In addition to the competent execution of the respective tasks, other factors influence the success of the project up to the entire Six Sigma Program of a company. Leyendecker et al. (2011) used the Six Sigma methods to investigate the influences on the success of Six Sigma projects and programs in companies (Fig. 14.5).

The **suitability of** the **certification projects** significantly **determines the success of these projects** at $R^2 = 35\%$ ($p < 0.5$). Therefore, we check Project-Topics very carefully for their suitability.

And just the success of the **certification projects determines the target achievement of subsequent projects** at $R^2 = 12\%$ ($p < 0.5$). Additionally, $R^2 = 34\%$ ($p < 0.5$) of the **target achievement** of all projects is **determined by the activity level** of the Six Sigma program.

It is often recommended to give the Belts about 10% of their working time for Six Sigma projects. In practice, however, only 26% of the companies surveyed follow this recommendation. Their Green Belts **complete all their projects in addition to**

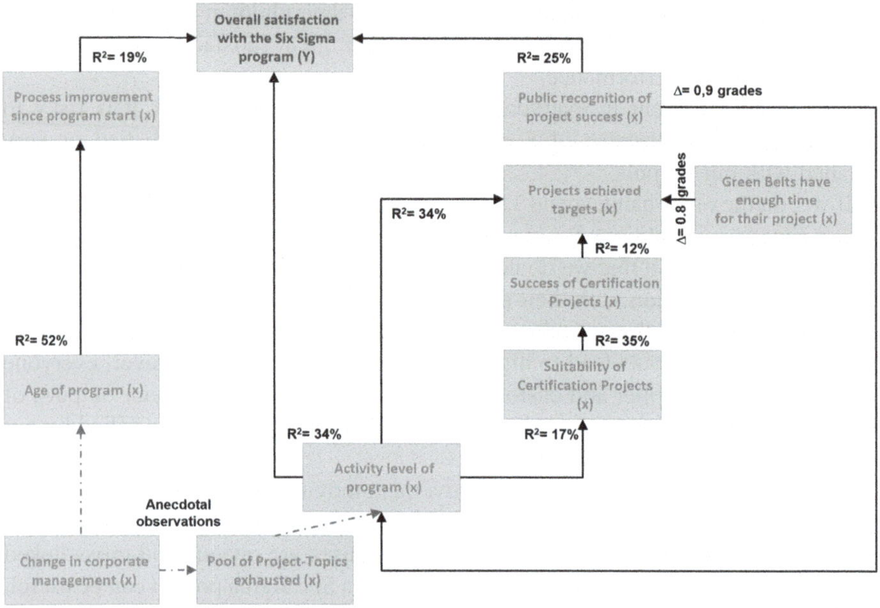

Fig. 14.5 Determinants of successful projects and the overall success of Six Sigma programs in German companies ($N = 78$, new results based on Schindewolf's (2011) data)

their daily work. This **time burden has no influence on the success of the certification projects**. However, the **time pressure affects the achievement of targets in subsequent projects is negatively** ($p < 0.1$, 0.8 grades difference on a 10-level scale).

Although the companies differ in the way they reward the achievement of project targets, most companies prefer the: *Personal recognition*, e.g., by the Sponsor/process owner (Fig. 14.6).

None of the recognitions provides an additional contribution to achieve the project targets, not even: *Financial boni*. However the: ***Public recognition*** of the project performance, e.g., by internal media and appropriate events, is a significant **indicator of the activity level** of a Six Sigma program ($p < 0.5$, 0.9 grades on a 10-level scale). And an **active Six Sigma program determines** the **target achievement** of the projects at $R^2 = 34\%$ ($p < 0.5$) **and** also the **overall satisfaction** with the program at $R^2 = 34\%$ ($p < 0.5$).

The overall satisfaction with the program is not determined by the age of the company program. The satisfaction with the program, therefore, does not change with the learning curve in dealing with Six Sigma. But the **age of the program exponentially determines** the overall **improvements of the company process** at

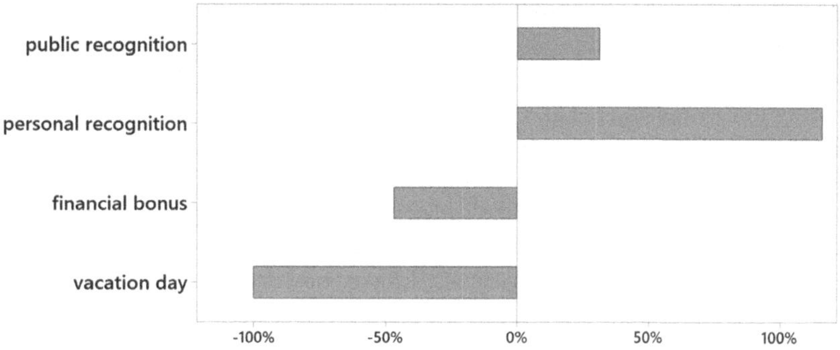

Fig. 14.6 Significant differences between rewards for successful Six Sigma projects (targets achieved) in German companies. Positive percentage values (personal recognition, public recognition) indicate rewards that are granted relatively more frequently, negative percentages indicate significantly fewer granted rewards (financial bonus, vacation day). (Chi^2 Goodness-of-Fit-Test, $p < 0.001$, $N = 78$, new results based on Schindewolf's (2011) data)

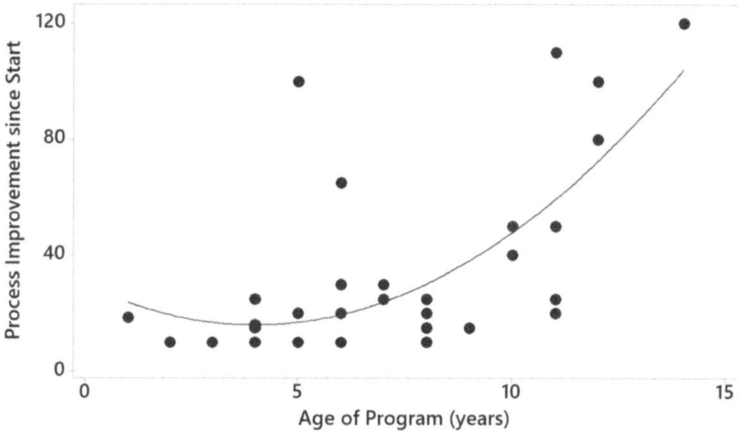

Fig. 14.7 The age of the Six Sigma program (x) exponentially determines the overall process improvements in German companies (Y) to $R^2 = 52\%$ ($N = 78$, $p < 0.5$, new results based on Schindewolf's (2011) data)

$R^2 = 52\%$ ($p < 0.5$). And these process improvements determine the overall satisfaction at $R^2 = 19\%$ ($p < 0.5$) (Fig. 14.7).

It is almost trivial to say that the age of the program also determines its life span. But because Six Sigma is based on scientific methods, it can evolve accordingly and remain up to date. However, there are a few factors that can lead to disappointment and eventually end the program.

14.9 Dissatisfaction with Six Sigma and Termination of Programs

According to Leyendecker et al. (2011), about 10% of employees in companies with established Six Sigma programs (age > 6 years) had completed Six Sigma training. **Black Belts** were **1%** of employees **4.8%** were **Green Belts.** This is a big investment and requires a ROI through the financial benefits of the projects.

Primarily the statistics of Six Sigma is demanding. And every certified Green Belt can be proud if these tools are mastered. I have, however, repeatedly experienced some **Belts treating their colleagues arrogantly**, especially members of Continuous Improvement Teams. This is harmful not just to Six Sigma. I know that Six Sigma and Lean-Thinking are two opposite sides of the same coin and that processes can be optimized with this currency. Six Sigma and Lean methods draw their justification from their strengths but are weak on the other side of the coin (Sect. 13.2). There is no reason for a Belt to be arrogant.

Some Belts and Sponsors were frustrated about the effort that a Six Sigma project requires. And some said that the problems could have been solved simply by common sense. I cannot prove this afterward. But for this reason, the Project-Definition already asks for solution ideas and you should definitely implement these ideas, if they will solve the identified process problems (Sect. 4.3.4). The Six Sigma method organizes and details common sense and the (data based) tools prevent wrong conclusions for problems for which there is no valid solution yet.

Six Sigma was in the past sometimes **praised as a superior saviour**. **This raised high expectations**, which were disappointed by the practice. Leyendecker et al. (2011) compared the **expected benefits of managers/Sponsors at the start** of their Six Sigma program **with** their **assessments of the overall achieved benefits** at the time of the survey and differentiated between for companies with a program age <= 6 years (t_1 start-up) and a program age > 6 years (t_2 established), regarding the criteria (Fig. 14.8):

1. Customer Complaints
2. Customer Retention
3. Revenue
4. Costs
5. Process-Capability (in general)
6. Quality (defects/suitability)
7. Availability (time/volume)
8. Consumption (input/resources)
9. Development of Products

The results of the survey show that there is:

1. (a) No significant difference in the expected benefit between the criteria and (b) No significant difference in the achieved benefits between the criteria. According to the relationships between problems (quality, availability,

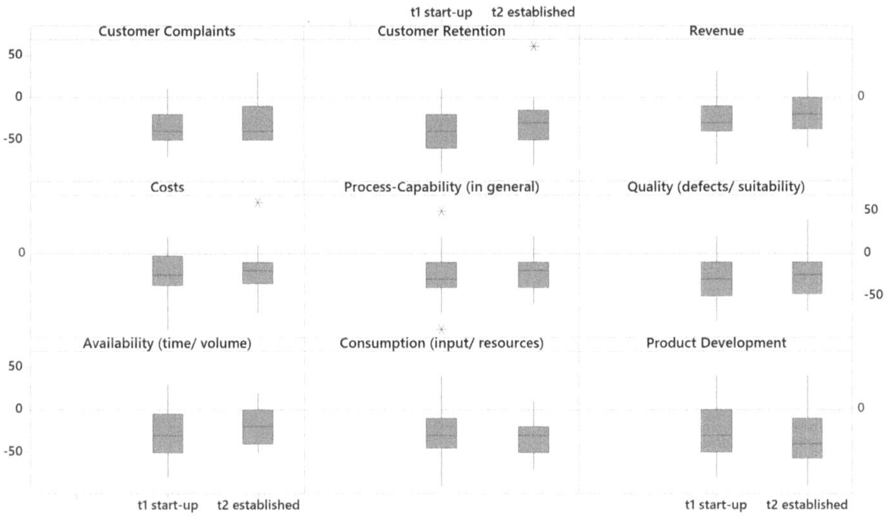

Fig. 14.8 Achieved benefits minus expected benefits for 9 Six Sigma success criteria differentiated between companies with a young Six Sigma program (t_1 start-up, age <= 6 years) and companies with an established Six Sigma program (t_2 established, age > 6 years) ($N = 78$, new results based on Schindewolf's (2011) data)

consumption) and their effects (customer satisfaction, costs), the result corresponds to our assumptions (#6): There is a clear relationship, e.g., between solving problems and reducing costs: If managers/Sponsors were (dis)satisfied with the problem-solving, then they were also (dis)satisfied with the effects on costs and customer.

2. The expected benefit at the start of the program was for each of the criteria higher than the achieved benefit at the time of the survey. This applies to young programs (age <= 6 years) as well as to mature programs (age > 6 years). Although none of these differences between expected and achieved benefits were statistically significant, they are consistent across all criteria – more expected than achieved.

Therefore, I would like to ask you **not to formulate your expectations too optimistically** in order to protect you from disappointments. Six Sigma is not a saviour.

I remember a threatening situation of a trading house with a very long tradition. **Shortly before the insolvency**, a Six Sigma program was to provide the upswing. However, none of the potential Sponsors were willing to define problems in their area of responsibility. They feared to uncover their own past failures and be held responsible for the insolvency. Half a year later, the company went bankrupt. **It is too late for Six Sigma in a heavy storm.**

A bad time for the introduction of Six Sigma is also the times of **restructuring** and times before **mergers and acquisitions**. In times of transformation the

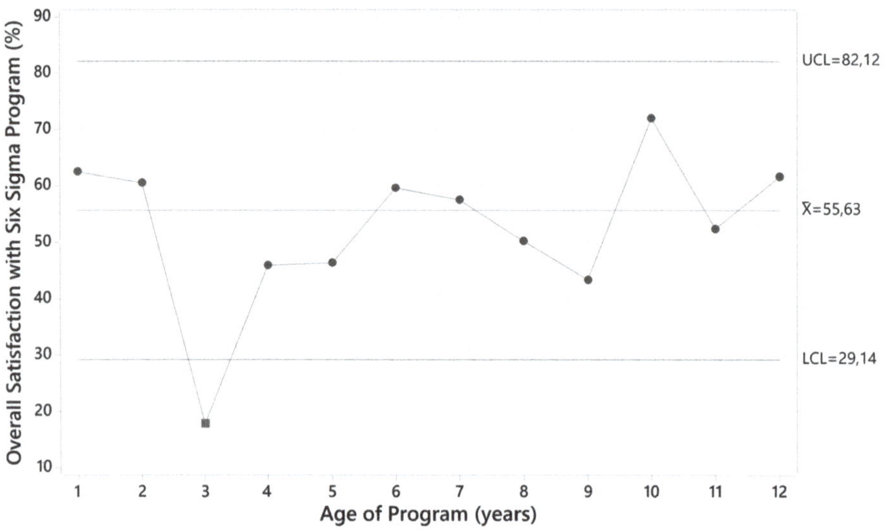

Fig. 14.9 Overall satisfaction of managers/Sponsors in German companies with their Six Sigma programs of different ages ($N = 78$, new results based on Schindewolf's (2011) data)

employees are insecure, the processes are unstable. However, Six Sigma is very well suited for the design and optimization of new processes after restructuring.

Despite the differences between the expectations on a Six Sigma program and practical experience, at the time of the study (2011) there were companies whose program was already 16 years old. I know some of these companies and they still practice Six Sigma today. However, the time course of the overall satisfaction of the companies with the program shows a threatening phase in the third year after the start, before the satisfaction stabilizes again on a relatively constant level (Fig. 14.9).

What are the causes of this satisfaction drop?

1. For companies that prematurely terminate Six Sigma, **Green Belts on average completed only 1.2 Six Sigma projects**, including the certification project. In these companies only one of five certified Green Belts carries out another project. For the company, the effort for the certification was not worthwhile. And of course the management recognizes this, terminates the program, and avoids further risks, although the Six Sigma resources remain wasted in the future.
2. The **time burden** of a Six Sigma project in addition to the daily work is an unattractive challenge for many Belts to take on further projects after certification (see Fig. 14.5).
3. Many **Green Belts lack the recognition/incentive** to carry out another Six Sigma project in addition to their daily work or feel insufficiently supported by the Sponsor.
4. In some cases, **unsuitable or too complex Project-Topics were selected** for these Green Belts and the projects were then terminated.

5. In some companies **two different managers are responsible for Six Sigma and Lean-Thinking**. In the worst case, this leads to murderous competition between two sides of the same coin. Successful **Lean-Thinking and Six Sigma programs** are the **sole responsibility of one person** reporting to the CEO.
6. If the focus is only on strategic relevant topics, some managers are **convinced** after 3 to 5 years **there is nothing left to improve** in the processes/products. This resembles an athlete, who terminates the training after 3–5 years and is convinced to qualify for the next competitions.
7. The **top management is replaced** and demands a **new start** in many areas.

There is nothing you can do about the seventh cause but present the track record to the new CEO. However, as a Sponsor you can help to eliminate, adjust, or circumvent all other causes by your solution ideas. At least that would be the Six Sigma approach. There are rational reasons why companies whose programs have lasted for more than 20 years are still identifying suitable topics and implementing projects with the greatest economic benefit. These are the Six Sigma experts. Both Six Sigma and Lean-Thinking programs benefit from the sustainable demand for improvement and the resulting culture of change. And I am convinced that there is no more efficient and effective method for complex problems for which data is available or can be collected.

References

Leyendecker, B., Schindewolf, S., Hutwelker, R., & Weigel, H. (2011, July). Erfolgsfaktoren für die Etablierung von Six Sigma. *QZ—Qualität und Zuverlässigkeit*.

Schindewolf, S. (2011). *Empirische Untersuchung des Erfolgs und der Erfolgsfaktoren bei der Ein- und Fortführung von Unternehmensprogrammen am Beispiel Six Sigma*. Unpublished Master Thesis, Hochschule Koblenz.

Notes for Six Sigma Experts on the Concept of this Course 15

Six Sigma has evolved continuously since the first draft by Bill Smith and Mikel Harry at Motorola (1988). And there is a common understanding among experts that the DMAIC is not an ideology but a scientific empirical method. This allows to assign any useful tool to the DMAIC. There is also large convergence in the published tool-lists. However, the literature and discussions with colleagues also show differences in the selection, phase assignment, sequence, scope, purpose, and design of the tools. In my experience, Belt candidates are confused when confronted with lists of 100 DMAIC tools, for example. They remind me of a walk through a DIY store: a treasure trove for the expert and an overload for the layman. To provide orientation for Belt candidates in particular, I have pursued three goals with the selection, sequence, and modification of the tools:

1. Offer one standard tool-set suitable for all business processes in production and service
2. Cover the entire project cycle from the identification of a problem to the evaluation of its solution, except for statistics
3. Link the tools chronologically and integrate them seamlessly so that the output of one tool serves as input for the following tool and the flow of information in the project can be monitored.

15.1 Standard Tool-Set for all Business Processes

The methodology in a Six Sigma project basically follows scientific principles (Fig. 2.13):

- Differences and relationships between attributes of objects are observed in a section of reality (maxi-cookies vs. mini-cookies; oven temperature and cookie texture).

© Springer Nature Switzerland AG 2019

R. Hutwelker, *Six Sigma Green Belt Certification Project*, Management for Professionals, https://doi.org/10.1007/978-3-030-31915-1_15

- The observed differences and relationships are mapped as assumptions in a descriptive and explanatory model. Six Sigma is based on an Input–Process–Output (IPO) model and a problem–solution cycle, standardized in the DMAIC.
- Data is collected for the attributes of the objects and the assumptions are tested statistically. The results confirm or contradict the assumptions about section of reality.

In developing a tool standard for projects within this framework, the claim of its general validity competes with its concrete practical relevance. This trade-off is balanced here with a universally valid but more differentiated IPO model, combined with a universally valid but more differentiated problem-solving cycle.

The integrated process-problem-solving model is the basis of my approach (Sects. 2.4.2 and 2.4.3). It separates between *input*, *activity*, and *output*. The output is always a countable material or immaterial *object* (cookie; data-field) with *attributes* (taste; entry), that more or less important for the customer and management.

A *problem* is defined as the deviation of the current status of a repeated created and thus countable object (input) from the expectation of this object, regarding an important attribute. Each of these attributes can be assigned to one of the categories: *Quality* (wrong, defect, bad, or unsuitable), *availability* (required quantity at the given time), and *consumption* (of input and resources). This focus on the status of certain *attributes* of certain *objects* determines a universal template for describing a *problem*: OBJECT ATTRIBUTE DEVIATION (COOKIE TASTE BAD, DATA-FIELD ENTRY WRONG).

The *activity* in the process *transforms* the input (sugar, flour,) to the output (cookie) or *transfers* the object from the input place to the output place. With this perspective, the *inputs* (shopping list, ingredients) of the activity can and need to be distinguished from the *methods* leading the activities (recipe, kitchen standards) as well as from the *resources* involved in the execution of the activity (Chef, kitchen scale, energy) for each process step.

In this process-model (Fig. 2.7) each *problem* of an output, i.e., each important deviation in on of its attributes, is caused exclusively by *negative influences* of inputs (x_i), methods (x_m), and resources (x_r) on the activities of this process.

These influences can be chronologically divided into *triggering influences*, as the last causes in a chain before the problem occurs and their *root causes* as the first causes of a chain. This incident-analytical perspective on the triggers in the activities leading to problems of the outputs (Fig. 2.8) focuses on the observable events and facilitates their operationalization as measurands.

Finally, problems concerning quality, availability, and consumption of outputs have effects on the *costs of the company* and the *satisfaction/loyalty of customer*. And *Solutions* of the *problems* and thus a relaxation of their *effects* focus on the elimination, circumvention, or adjustment of the *root causes*.

The grid of this process-problem-solving model helps the Belt to collect and structure information to monitor the results and to plan the next steps of the project.

15.2 Coverage of the Entire Project Cycle

It was important to me to have one set of necessary and sufficient tools for all Six Sigma projects, a backbone to which Belt candidates can develop an understanding of the Six Sigma logic in accordance with their common sense.

The selection, sequence, and modifications of tools in sigmaGuide and thus in this course depends on the purpose of the phases:

- The **DEFINE** phase deals with the problems (**Y**) of the outputs and the resulting effects (**Z**) on the costs and customer satisfaction.
- In the **MEASURE** phase it is first about the influences (**X**) in the input and the activities on the problems (**Y**), then about the relationships between influences (**x**) and problems (**Y**) and the derived hypotheses (**x–Y pairs**), then about the Data-Collection-Plan for **x** and **Y**.
- In the **ANALYSE** phase, the collected data are evaluated, with graphical analyses and Process-Capability indices (**Y**), with the statistical tests of the hypotheses (**x–Y pairs**) up to the deduction of root causes (**x′**) of the problems (**Y**).
- The root causes (**x′**) are the interfaces for solutions (**S**) in the **IMPROVE** phase.
- The impact of the solutions (**S**) on the problems (**Y**) and the resulting financial benefits (**Z**) is finally evaluated and sustainably ensured in the **CONTROL** phase.

The result is a set of tools, with a clear purpose and focus in the process-problem-solving model. And the purpose of the phases allows to arrange the tools chronologically, according to the principle: from the general to the specific. This assignment corresponds to the names of the phases, is consistent with scientific practice and understandable to Belt candidates.

Three tools (Fig. 15.1) are not primarily used for problem-solving but to organize and communicate the project and to assure **c**ooperation (OCC). With the Project-Charter an agreement is made between the Sponsor, Green Belt, and the team. The Stakeholder-Communication serves to assure the necessary support of persons who are affected by or involved in the project. And the Summary and Benefits gives an overview of the lessons learned and the financial and other benefits of the project.

The structure allows to select exactly the tools that are relevant for the focused variables (**Y–Z–X–S**) in the phases. This prevents, for example, FMEA (**X–Y–Z**) and Process-Mapping (**X**) from already being used in the DEFINE, the Hypotheses in the ANALYSE, and the Data-Collection-Plan (**X–Y**) in the late IMPROVE phase.

Of course, new Hypotheses can emerge from the Root-Cause-Analysis. These new Hypotheses would require a loop back to MEASURE so that the necessary data collection can also be planned and carried out. If, however, the formulation of Hypotheses is assigned to the ANALYSE phase, and if these Hypotheses contain variables for which no data was collected in MEASURE, then the loop back becomes the standard. That is neither efficient nor meaningful. And no scientist would collect data without an underlying hypothesis.

Phase	Tool	Purpose	Focus				
			S (Solutions for Causes)	X (Influences/ Causes)	Y (Problems/ Targets)	Z (Effect of Problems)	PM (Project Management)
D	Project-Topic	Identify a weakness of a process			Y	Z	
	Project-Definition	Collect available information on the weakness	S	X	Y	Z	
	SIPOC	Specify the project field		X	Y		
	Voice-to-Critical	Define the problem			Y		
	Project-Charter	Sign a project contract			Y	Z	PM
	Stakeholder-Communication	Plan the communication with the stakeholder					PM
M	Input-Analysis	Identify triggering influences of the input		X			
	Process-Mapping-Analysis	Identify triggering influences of methods and resources		X			
	C&E-Matrix	Formulate assumptions about problem-trigger relationships		X	Y		
	Data-Collection-Plan	Plan the data collection		X	Y		
	Hypotheses	Formulate hypotheses about problem-trigger relationships		X	Y		
		Collect data					
A	Statistics	Determine the process performance			Y		
	Statistics	Test hypotheses about trigger-problem relationships		X	Y		
	Hierarchy Tree	Identify the root causes for significant trigger		X	Y		
I	Solutions	Identify solutions to eliminate, adjust or circumvent the root causes	S	X			
	FMEA	Identify risks of the solutions	S	X	Y	Z	
	Action-Plan	Specify solutions as concrete work packages	S			Z	PM
		Implement the work packages and collect new data					
C	Statistics	Determine the new process performance			Y		
	Statistics	Test the before-after difference for all defined problems		X	Y		
	Process-Management-Plan	Develop a plan to monitoring and continuously improving the process		X	Y		PM
	Summary & Benefits	Demonstrate the financial and other benefits of the solutions	S	X	Y	Z	

Fig. 15.1 Sequence of DMAIC tools, their purpose, and focus within the process-problem-solving model

15.3 Chronological Linkage and Integration of Tools

In many lists, the tools are presented isolated from each other. Companies often use a bundle of separate Word and Excel templates representing the basic ideas of the tools. And the operating instructions for the application of these tools often leave many degrees of freedom for editing. As a result, some projects move away from their scope very quickly—others remain vague. Therefore, I added three new tools and modified some of the existing tools.

In the literature, the DEFINE phase typically begins with the problem definition followed by SIPOC. The exclusive focus on only selecting strategically relevant problems is a risk for the Six Sigma company program (Sects. 14.8 and 14.9). Furthermore, it has always been a challenge for me in the past to identify suitable project topics in meetings with the management. In contrast, I have always benefited from the employees' point of view. They know the strengths and weaknesses of their working environment. Thus, I introduced two tools at the start of the DMAIC in my sigmaGuide:

1. Project-Topic: Serves employees to identify process weaknesses, translate them into Six Sigma logic, and provide indications of their suitability as projects.
2. Project-Definition: Serves the management in selecting of the most promising topics for project.
3. Stakeholder-Communication: Serves the belt to identify influential but skeptical colleagues and win them to support the project.

The modifications of the tools concern the specification and quantification of their inputs and outputs in order to harmonize the conditions and consequences of the tools and their interfaces. This is just as true for information as it is for hardware, because otherwise the hammer will treat a screw as a nail.

The important inputs and outputs for the tools are specified by the Process-Problem-Solving Model (Sects. 2.4.2 and 2.4.3) and the Root-Cause-Analysis (Sect. 8.5) indicating how a problem (Y) can be defined and how its triggering influences and their root causes (x; x') can be identified.

4. SIPOC: Strict discrimination of the countable objects of input and output, supplemented by the new attributes added in each process step, from the activities of the process.
5. Voice-to-Critical: Integration of the tools: Voice of Customer/Business, Critical Customer/Business Requirements and the Kano-Model to specify the problems (Y) of the SIPOC outputs as objects with deviating attributes. By operationalizing the three Kano categories as characteristics of different power functions, together with the estimated degree of the requirement fulfillment, the *severity* is calculated for each problem. Problems with the highest severity are CtQs.
6. Input-Analysis: Specification of inputs as objects with deviating attributes supplemented by the estimated *frequency* of the deviation, serving as triggering influences (x_i) in the model.
7. Process-Mapping/-Analysis: Specification of the guiding methods and executing resources and their negative influences on the activities of the process, supplemented by the estimated *frequency* of their occurrence, serving as triggering influences (x_m; x_r) in the model.
8. C&E Matrix: Matching of the triggering influences of the inputs and activities with their related frequencies (x) and problems with their severity (Y). By evaluating the relationship strength of the resulting x–Y pairs, it is possible not only to calculate the total strength of each influence (x) on all problems (Y), but also the extent to which each problem (Y) is determined by all influences (x). In addition, the *risk* of each x–Y pair can be calculated as:

   Frequency of the influence (x) $*$ severity of the problem (Y) $*$ strength of the x–Y relationship. These quantified risks for all x–Y pair are shown in the chart: C&E Heatmap and form the assumptions for the following hypotheses.
9. Data-Collection-Plan: The attributes of influences and problems determine the appropriate statistical *scale-levels* of the measurands for x and Y. The distinction between *data* and underlying *scales* is necessary, because, e.g., the data [0; 1] can represent discrete attributes on a nominal scale, can indicate a ranking of attributes on an ordinal scale, can represent discrete and even continuous attributes with their values on a cardinal scale. The knowledge of the scale level however determines the appropriate statistical parameters for the data, appropriate charts, performance indicators, and statistical tests.
10. Hypotheses: Determining the scale levels for x and Y and calculating the risk for each x–Y pair allows Hypotheses to be automatically generated, sorted by their importance, and allows to propose appropriate statistical tests for each x–Y pair.
11. In the IMPROVE phase, the tools for Solution finding, the FMEA and the measures of the Action-List are interlinked.

12. Finally, the Process-Management-Plan in CONTROL uses the variables of the Data-Collection-Plan as input.

These modifications allow to sequence and seamlessly integrate the tools from the identification of a Project-Topic at the beginning of DEFINE to the formulation of Hypotheses at the end of MEASURE and to continue the DMAIC after the statistical tests for the IMPROVE and CONTROL phase. Therefore, every candidate of our certification projects knows the answers to the following questions:

1. Which tools are mandatory for (almost) every project?
2. In which order should the tools they be applied?
3. What information should be collected for a certain tool?
4. What is the output of a certain tool?
5. How can I monitor the flow of information from tool to tool?

These are the questions that motivated me to develop these tools, this course and to write this book. The biggest motivation, however, was to leave a testimony from me to my daughter Nora.

Answers to Questions

Video 4/ Ch. 2.2/ #1 (SC) **Please assign the roles to their explicit/implicit tasks in a Six Sigma program.**

		Identify project topics	Select projects	Support projects	Lead projects
1	Black Belts				x
2	Customer	x			
3	Employees	x			
4	Green Belts				x
5	Manager		x		
6	Master Black Belts		x		
7	Sponsors			x	
8	Subject Matter Experts			x	
9	Yellow Belts			x	
9	points				

Video 6/ Ch. 2.4.1/ #1 (SC) **Please assign the tasks of a Six Sigma project to it's DMAIC phases.**

		DEFINE	MEASURE	ANALSYE	IMPROVE	CONTROL
1	Identify project topic	x				
2	Calculate benefits of the project					x
3	Collect data		x			
4	Determine the problems (Y)	x				
5	Develop a process-management-plan					x
6	Develop solutions				x	
7	Formulate hypotheses		x			
8	Identify the influences (x)		x			
9	Identify the risks of measures				x	
10	Identify the root causes (x')			x		
11	Identify requirements on the product/ service	x				
12	Implement concrete measures				x	
13	Listen to the Voice of Customer	x				
14	Test hypotheses			x		
14	points					

Video 7/ Ch. 2.4.2/ #1 (SC) **In which part of a process can influences/causes (x) occur and in which do problems (Y) occur?**

		Input	Method	Activity	Ressources	Output
1	Influences (x)/ Causes (x')	x	x	x	x	
2	Problems (Y)					x
2	points					

© Springer Nature Switzerland AG 2019

R. Hutwelker, *Six Sigma Green Belt Certification Project*, Management for Professionals, https://doi.org/10.1007/978-3-030-31915-1

Video 7/ Ch. 2.4.2/ #2 (SC) **To which part of a process do these components of our bakery belong?**

		Input	Method	Activity	Ressources	Output
1	Vanilla	x				
2	Energy				x	
3	Write shopping list			x		
4	Chef				x	
5	Cookie					x
6	Oven manual		x			
7	Enter the order			x		
8	Order	x				
9	Cookie delivery note					x
10	Hygiene regulations		x			
10	points					

Video 7/ Ch. 2.4.2/ #3 (MC) **What types of problems can an output have?**

		Quality	Costs	Satisfaction	Availability	Consumption
1	An Output can have a problem in terms of:	x			x	x
1	points					

Video 8/ Ch. 2.4.3/ #1 (SC) **Please classify these observations from our bakery into the categories of the process-problem-solving model.**

		Influence (x)	Problem (Y)	Effect (Z)	Solution (S)
1	Cookies surface broken		x		
2	Vanilla not available	x			
3	Cookie diameter too large		x		
4	Customer complaints			x	
5	Cookie invoice wrong		x		
6	Adhere to recipe specifications				x
7	Kitchen scale incorrectly calibrated	x			
8	Forbid nibbling on the dough				x
9	Egg-yolk/ egg-white separation laborious	x			
10	Decline in sales of cookies			x	
11	Change chocolate supplier				x
12	Energy costs too high			x	
12	points				

Video 9/ Ch. 2.5/ #1 (SC) **From which indicator can the standard deviation already be estimated, based on the curve of a normal distribution?**

1	Skewness	
2	Kurtosis	
3	Inflection points	x
4	Quartiles	
4	points	

Video 9/ Ch. 2.5/ #2 (MC) **How can the process capability index be improved?**

1	Centring of the process	x
2	Reducing the scatter of the process	x
3	Changing the specification limits	x
4	Eliminating outliers	x
4	points	

Video 9/ Ch. 2.5/ #3 (SC)	Which z-value on the x-axis of the normal distribution corresponds to the area proportion of 3.4 ppm?	
1	1	
2	1.5	
3	4.5	x
4	6	
4	points	

Video 10/ Ch. 2.6/ #1 (SC)	Typically, a Six Sigma project is used to investigate:		
		... influences (x) and problems (Y)	... problems (Y)
1	Relationships between ...	x	
2	Differences between ...		x
2	points		

Video 13/ Ch. 4.2/ #1 (SC)	Please assign these influences, types of problems and the degree of measurability to the different project types.			
		Basic Improvement	Lean Sigma	Six Sigma
1	Method wrong/ incomplete/ not available	x		
2	Resource defect/ not available	x		
3	Activity laborious/ inefficient/ repeated		x	
4	Output defect/ not available			x
5	Quality problem			x
6	Availability problem		x	
7	Consumption problem (waste)	x	x	
8	x and Y measurable			x
9	only Y measurable		x	
10	nothing measuable	x		
10	points			

Video 16/ Ch. 6.1/ #1 (MC)	What are the four most important purposes of the SIPOC?	
1	Determine the important (intermediate) outputs	x
2	Define the limits of the project field	x
2	Inspect the used methods and resources of the process steps	
3	Get an overview of the influences of the process	
4	Evaluate the outputs of the process steps	
5	Identify the most important activities of the process steps	
6	Make the core process steps transparent within the project field	x
7	Identify the interviewpartner for the VoC	x
7	points	

Video 16/ Ch. 6.1/ #2 (SC)	Which word types should be used to specify inputs, process steps and outputs?		
		noun	verb + noun
1	Inputs	x	
2	Process-Steps		x
3	Outputs	x	
3	points		

Video 16/ Ch. 6.1/ #3 (MC)	What should be additionally specified in brackets for each output if possible?	
1	The number of variants of a product/ service	
2	The problem of the output	
3	The last added/modified attribute of the output	x
4	The achieved development stage of the output	x
4	points	

Video 17/ Ch. 6.2/ #1 (SC)	Please bring the following steps to completing the tool: Voice-to-Criticals in their correct order.					
		1st step	2nd step	3rd step	4th step	5th step
1	Define Critical Customer/ Business Requirements (CCR/ CBR) on the attributes			x		
2	Determine the problem type as a quality, availability or consumption problem					x
3	Identify wishes and complaints in the Voice of Customer/ Business (VoC/ VoB)	x				
4	Define the critical deviation of an attribute as a problem of the output				x	
5	Derive important attributes of the output from the wishes and complaints		x			
5	points					

Video 17/ Ch. 6.2/ #2 (SC)	Please assign the appropriate Kano category to the reactions of a customer resulting from the fulfillment and non-fulfillment of the requirements.				
		angry if **not** fullfilled	**indifferent** if **not** fulfilled	**indifferent** if fulfilled	**excited** if fulfilled
1	must-be	x		x	
2	more-/ less -is-better	x			x
3	delighter		x		x
3	points				

Video 17/ Ch. 6.2/ #3 (MC)	What information is used to determine the critical problems, the CtQ's?	
1	CCR/ CBR	
2	Kano-category (must-be, more-/ less-is-better, delighter)	x
3	VoC/ VoB	
4	Extent to which requirements are met	x
5	Problem type (quality, availability or consumption)	
6	Calculation of the severity of the problems	x
7	Prioritization of the problems based on their severity	x
7	points	

Video 19/ Ch. 6.4/ #1 (SC)	Please assign the appropriate activities to the possible causes of resistance to your project.				
		Communicate	**Qualify**	**Motivate**	**Integrate**
1	Fear of making mistakes		x		
2	The need for improvements is not seen	x			
3	Actual/presumed group pressure against changes				x
4	Additional workload/ less influence expected			x	
4	points				

Video 19/ Ch. 6.4/ #2 (SC)	Please assign the stakeholders to the different types according to their interest and power to support your project.				
		Defender	**Apathetics**	**Latents**	**Promotors**
1	high interest and high power				x
2	low interest and high power			x	
3	high interest and low power	x			
4	low interest and low power		x		
4	points				

Video 21/ Ch. 7.2/ #1 (MC)	What should you do if you could foresee that in your project only influences of the inputs trigger the defined problems?	
1	That's normal, no need to change anything	
2	I invite the supplier to join the project	
3	I shift the project field upstream so that the inputs become the outputs	x
4	I inform my sponsor	x
4	points	

Video 22/ Ch. 7.3/ #1 (SC)	Please bring the following steps to completing the tool: Voice-to-Criticals in their correct order.					
		1st step	2nd step	3rd step	4th step	5th step
1	What are the inputs and outputs of the activities?			x		
2	Which departments are involved?	x				
3	Which influences of methods and resources trigger the problems in which frequency?					x
4	Which specific activities are carried out?		x			
5	What are the methods and resources of the activities?				x	
5	points					

Video 22/ Ch. 7.3/ #2 (MC)	What is the purpose of the Process-Mapping/ -Analysis?	
1	Creation of a certifiable document	
2	Identify the root causes of the problems	
3	Identify triggers of the problems	x
4	Get an overview of the involved departments	
5	Identify loops, critical interfaces and unnecessary responsibility changes	x
5	points	

Video 23/ Ch. 7.4/ #1 (SC)	In the C&E Matrix, influences (x) in the process are matched against problems in outputs (Y).			
		Relationship between x and Y	**Overall influence of each x**	**Overall determination of each Y**
1	What is evaluated in the cells of the C&E Matrix?	x		
2	What is the result per row?		x	
3	What is the result per column?			x
3	points			

Video 24/ Ch. 7.5/ #1 (SC)	Please assign the following measurands to their underlying scale levels.			
		nominal	ordinal	cardinal
1	Cycle time			x
2	Different types of waste	x		
3	Weight in grams			x
4	Variants of a product	x		
5	School grades		x	
6	Length in millimetres			x
7	Defect types	x		
8	Costs			x
9	Satisfaction levels		x	
10	Human littering types			
10	points			

Video 24/ Ch. 7.5/ #2 (MC) **Please assign the following parameters to the correct scale levels and determine the respective parameter types.**

		nominal	ordinal	cardinal	Position Parameter	Scattering Parameter
1	Mean			x	x	
2	Percentile		x			x
3	Range		x			x
4	Mode	x			x	
5	Median		x		x	
6	Standard deviation			x		x
7	Interquartile range		x			x
8	Variance			x		x
8	points					

Video 25/ Ch. 7.6/ #1 (SC) **Which hypothesis is ...**

		H0	HA
1	... tested in a statistical test?	x	
2	... accepted if p < alpha?		x
3	... accepted if p >= alpha?	x	
3	points		

Video 25/ Ch. 7.6/ #2 (MC) **Which scale levels of x and Y define the type of content hypothesis?**

		x: nominal	x: cardinal	Y: nominal	Y: cardinal
1	Difference Hypothesis	x			x
2	Relationship Hypothesis		x	x	x
2	points				

Video 27/ Ch. 8.2/ #1 (SC) **Please assign the following diagrams to the lowest necessary scale level.**

		nominal	ordinal	cardinal
1	Histogramm			x
2	Box Plot		x	
3	Time Series Plot		x	
4	Pareto Chart	x		
5	Scatter Plot			x
6	Interval Plot			x
6	points			

Video 28/ Ch. 8.3/ #1 (SC) **Please assign the following control-charts to the required scale levels.**

		nominal	ordinal	cardinal
1	u-Chart	x		
2	xbar-S-Chart			x
3	xbar-R-Chart		x	
4	I-MR-Chart			x
5	p-Chart	x		
5	points			

Video 28/ Ch. 8.3/ #2 (SC)	Please assign the following capability indices to the required scale levels.			
		nominal	ordinal	cardinal
1	Defect-%	x		
2	Cp/ Cpk			x
3	Pp/ Ppk			x
4	Defects per Unit	x		
5	DPMO	x		
5	points			

Video 29/ Ch. 8.4/ #1 (MC)	Please assign the required scale levels for x and Y to the statistical tests.					
		x: nominal/ 2 levels	x: nominal/ > 2 levels	x: cardinal	Y: nominal	Y: cardinal
1	t-test	x				x
2	Chi-Squared Test	x	x		x	
3	Correlation			x		x
4	ANOVA		x			x
5	Regression-Analysis			x		x
5	points					

Video 29/ Ch. 8.4/ #2 (SC)	Which alpha levels are suitable for which type of examination?			
		alpha= 10%	alpha= 5%	alpha= 1%
1	Preliminary investigation of an assumption	x		
2	Standard decision		x	
3	Major decision			x
3	points			

Video 29/ Ch. 8.4/ #3 (SC)	Which of these statistical test data determines the ...					
		p-value	alpha	difference between means	level of correlation	level of determination
1	statistical significance	x	x			
2	practical relevance			x	x	x
2	points					

Video 30/ Ch. 8.5.1/ #1 (MC)	Which statement apply to the prioritization of problems?	
1	The Allows Root-Cause-Analysis to begin with the most important problem	x
2	Dependent problems may not need to be solved	x
3	The root problems often occur less frequently than the common problems	x
4	Prioritization can be based on frequencies and logical dependencies	x
5	Logical and frequency-based prioritization leads to identical results	
6	In the case of dependent problems, the Root-Cause-Analysis may be omitted	x
7	Root-Cause-Analysis should always start with the problem with the highest non-conformity	
7	points	

Video 31/ Ch. 8.5.2/ #1 (SC)	Which attributes do the Fisbone/Ishikawa diagram and the hierarchy tree have?		
		Fishbone Diagram	Hierarchy Tree
1	Collect information on all influence areas	x	
2	Identifies the trigger		x
3	Collects sometimes irrelevant information	x	
4	Uses the associative capabilities of experts	x	
5	Forces to detect chronological dependencies		x
6	Allows to integrate statistical tests		x
6	points		

Video 31/ Ch. 8.5.2/ #2 (SC)	What logical relationships do these statistical tests represent?		
		If ... then ...	The ... the ...
1	Chi-Squared Test	x	
2	Correlation		x
3	Regression-Analysis		x
3	points		

Video 31/ Ch. 8.5.2/ #3 (SC)	What does the word: not or the prefix: un- often indicate in a Root-Cause-Analysis?	
1	Root-Cause	
2	Trigger	
3	Problem	
4	Effect	
5	Solution	x
5	points	

Video 33/ Ch. 9.2/ #1 (MC) **Which of these methods support the elimination, circumvention or adjustment of root causes?**

		6-3-5 Method	Six Thinking Hats	Design of Experiments (DoE)	TRIZ	Morphological Analysis
1	Elimination	x	x		x	x
2	Circumvention	x	x		x	x
3	Adjustment			x		
3	points					

Video 34/ Ch. 9.3/ #1 (SC) **Please assign the evaluation categories to the risk components of the FMEA.**

		Detectability	Severity	Probability
1	Cause			x
2	Problem/ Defect	x		
3	Effect		x	
3	points			

Video 37/ Ch. 10.3/ #1 (MC) **What do the tests of control-charts indicate?**

1	Outlier	x
2	Shifts	x
3	Drifts	x
4	Relationships	
5	Cycles	x
6	Differences	
7	Oscillations	x
8	Significances	
9	Trends	x
10	Capabilities	
10	points	

Video 37/ Ch. 10.3/ #2 (MC) **Please assign the the necessary requirements to the following capability indices and control-charts.**

		Data normal distributed	N ≤ 8	N > 8
1	I-MR-Chart	x		x
2	Cp/ Cpk	x		x
3	Pp/ Ppk	x		x
4	xbar-S-Chart	x		x
5	xbar-R-Chart		x	
5	points			

Video 37/ Ch. 10.3/ #3 (SC) **What do these capability indices evaluate?**

		Cpk/ Ppk/ Z.bench	Defect %/ Z.bench	DPU
1	Tolerance/ standard deviation	x		
2	Defects outside specification		x	
3	Defects per Unit			x
3	points			

Video 38/ Ch. 10.4/ #1 (MC)	Which tests are suitable for a before-after comparison?	
1	2-Sample-t-Test	x
2	ANOVA	x
3	Chi-Squared-Test	x
4	Regression-Analysis	
5	Correlation	
6	2-Sample-%-Defective-Test	x
6	points	

Video 38/ Ch. 10.4/ #2 (SC)	Which result from a statistical test should be the basis for a conservative estimate of the financial benefit?	
1	Lower limit of the confidence interval	x
2	Expected value (e.g. the mean difference)	
3	Upper limit of the confidence interval	
3	points	

Video 39/ Ch. 10.5/ #1 (SC)	In which sequence can ambitious control limits for checking future data in a new process be determined?					
		1. Step	2. Step	3. Step	4. Step	5. Step
1	Estimate control limits from the subsample		x			
2	Exclude (at least) extreme values from calculation of LCL/ UCL			x		
3	Extract a subsamble of the new measurement data that is/ seems under control	x				
4	Create a control-chart for the complete sample of new measured data					x
6	Take the calculated LCL/ UCL as reference values for the final control-chart				x	
6	points					
235	**total score**					

Further Reading

Breyfogle, F. W. (2003). *Implementing Six Sigma*. Hoboken, NJ: Wiley.
Brook, Q. (2014). *Lean Six Sigma & Minitab*. Hampshire: Opex Resources.
Harry, M., & Schroeder, R. (2006). *Six Sigma: The breakthrough management strategy revolutionizing the World's top corporations*. Doubleday, NY: Currency.
Montgomery, D. C. (2012). *Design and analysis of experiments*. Hoboken, NJ: Wiley.
Staudter, C., Mollenhauer, J.-P., Meran, R., Roenpage, O., von Hugo, C., & Hamalides, A. (2003). *Design for Six Sigma + Lean Toolset*. Berlin: Springer.

Links

Certification: TU München School of Management: www.eec.wi.tum.de.
Costs of littering: Federal Office for the Environment/Bundesamt für Umwelt BAFU. (2011). https://www.igsu.ch/files/bafu_litteringkosten_2011.pdf (contains English Translation).
Documentation of selected garbage littered places in your project: World Cleanup App. https://www.worldcleanupday.org/map-it/?_branch_match_id=565815979123652917
Handouts and software also available at: www.sigmalogic.de
Six Sigma Yellow and Green Belt course: *Six Sigma and Lean: Quantitative Tools for Quality and Productivity (QPLSx)*: www.edx.org
The videos are also available at: https://www.youtube.com/channel/UCShsy50l5dFsoSXU__hBKSA
Worldwide waste statistics and initiatives: World Waste Platform. https://opendata.letsdoitworld.org/#/download

© Springer Nature Switzerland AG 2019

R. Hutwelker, *Six Sigma Green Belt Certification Project*, Management for Professionals, https://doi.org/10.1007/978-3-030-31915-1

The manufacturer's authorised representative in the EU is Springer Nature Customer Service Centre GmbH, Europaplatz 3, 69115 Heidelberg, Germany. If you have any concerns regarding our products, please contact ProductSafety@springernature.com

Printed and bound by CPI Group (UK) Ltd, Croydon, CR0 4YY
30/09/2026
02225650-0008